Handbook of X-ray Astronomy

Modern X-ray data, available through online archives, are important for many astronomical topics. However, using these data requires specialized techniques and software. Written for graduate students, professional astronomers, and researchers who want to start working in this field, this book is a practical guide to X-ray astronomy.

The handbook begins with X-ray optics, basic detector physics, and charge-coupled devices, before focusing on data analysis. It introduces the reduction and calibration of X-ray data, scientific analysis, archives, statistical issues, and the particular problems of highly extended sources. The book describes the main hardware used in X-ray astronomy, emphasizing the implications for data analysis. The concepts behind common X-ray astronomy data analysis software are explained. The appendices present reference material often required during data analysis.

KEITH ARNAUD is at the Center for Research Excellence in Space Science and Technology, NASA Goddard Space Flight Center, and is an Associate Research Scientist in the Astronomy Department, University of Maryland. A veteran of X-ray astronomy, he is recognized worldwide as an expert on data analysis techniques.

RANDALL SMITH is an astrophysicist in the High Energy Astrophysics Division of the Smithsonian Astrophysical Observatory. He is internationally known for his work on spectral emission from astrophysical plasmas and the underlying issues of atomic physics.

ANETA SIEMIGINOWSKA is an astrophysicist in the High Energy Astrophysics Division of the Smithsonian Astrophysical Observatory. She has worked in both theoretical and observational aspects of X-ray astronomy with interests in extragalactic radio sources, quasars, powerful jets, and statistical methods.

Cambridge Observing Handbooks for Research Astronomy

Today's professional astronomers must be able to adapt to use telescopes and interpret data at all wavelengths. This series is designed to provide them with a collection of concise, self-contained handbooks, which covers the basic principles peculiar to observing in a particular spectra region, or to using a special technique or type of instrument. The books can be used as an introduction to the subject and has a handy reference for use at the telescope, or in the office.

Series editors

Professor Richard Ellis, Department of Astronomy, *California Institute of Technology*
Professor Steve Kahn, Department of Physics, *Stanford University*
Professor George Rieke, Steward Observatory, *University of Arizona*, Tucson
Dr Peter B. Stetson, Herzberg Institute of Astrophysics, *Dominion Astrophysical Observatory*, Victoria, British Columbia

Books currently available in this series:

1. Handbook of Infrared Astronomy
   I. S. Glass

3. Practical Statistics for Astronomers
   J. V. Wall, C. R. Jenkins

4. Handbook of Pulsar Astronomy
   D. R. Lorimer, M. Kramer

5. Handbook of CCD Astronomy, Second Edition
   Steve B. Howell

6. Introduction to Astronomical Photometry, Second Edition
   Edwin Budding, Osman Demircan

7. Handbook of X-ray Astronomy
   Edited by Keith Arnaud, Randall Smith, and Aneta Siemiginowska

Handbook of X-ray Astronomy

Edited by

KEITH A. ARNAUD,[1,2] RANDALL K. SMITH,[3] AND
ANETA SIEMIGINOWSKA[3]

1. CRESST, NASA Goddard Space Flight Center
2. Astronomy Department, University of Maryland
3. Smithsonian Astrophysical Observatory

CAMBRIDGE
UNIVERSITY PRESS

University Printing House, Cambridge CB2 8BS, United Kingdom

One Liberty Plaza, 20th Floor, New York, NY 10006, USA

477 Williamstown Road, Port Melbourne, VIC 3207, Australia

314-321, 3rd Floor, Plot 3, Splendor Forum, Jasola District Centre, New Delhi - 110025, India

79 Anson Road, #06-04/06, Singapore 079906

Cambridge University Press is part of the University of Cambridge.

It furthers the University's mission by disseminating knowledge in the pursuit of education, learning and research at the highest international levels of excellence.

www.cambridge.org
Information on this title: www.cambridge.org/9780521883733

© Cambridge University Press 2011

First published 2011

A catalogue record for this publication is available from the British Library

Library of Congress Cataloging in Publication data
Handbook of X-ray astronomy / edited by Keith A. Arnaud,
Randall K. Smith, Aneta Siemiginowska.
p.   cm. – (Cambridge observing handbooks for research astronomers)
Includes index.
ISBN 978-0-521-88373-3 (hardback)
1. X-ray astronomy.   I. Arnaud, Keith A., 1959–   II. Smith, Randall K.
(Randall Knowles), 1969–   III. Siemiginowska, Aneta.
IV. Title.   V. Series.
QB472.H36   2011
522′.6863 – dc23   2011023034

ISBN 978-0-521-88373-3   Hardback

Contents

Contributors

Daniel A. Schwartz, Harvard-Smithsonian Center for Astrophysics, 60 Garden Street, Cambridge, USA, MA 02138

Richard J. Edgar, Harvard-Smithsonian Center for Astrophysics, 60 Garden Street, Cambridge, USA, MA 02138

Catherine E. Grant, MIT Kavli Institute for Astrophysics and Space Research, 77 Massachusetts Avenue, Cambridge, USA, MA 02139

Keith A. Arnaud, NASA Goddard Space Flight Center, Code 662, Greenbelt, USA, MD 20771

Randall K. Smith, Harvard-Smithsonian Center for Astrophysics, 60 Garden Street, Cambridge, USA, MA 02138

Aneta Siemiginowska, Harvard-Smithsonian Center for Astrophysics, 60 Garden Street, Cambridge, USA, MA 02138

K. D. Kuntz, JHU Department of Physics & Astronomy, 3400 N. Charles Street, Baltimore, USA, MD 21218

Introduction

X-ray astronomy was born in the aftermath of World War II as military rockets were repurposed to lift radiation detectors above the atmosphere for a few minutes at a time. These early flights detected and studied X-ray emission from the solar corona. The first sources beyond the Solar System were detected during a rocket flight in 1962 by a team headed by Riccardo Giacconi at American Science and Engineering, a company founded by physicists from MIT. The rocket used Geiger counters with a system designed to reduce non-X-ray backgrounds and collimators limiting the region of sky seen by the counters. As the rocket spun, the field of view (FOV) happened to pass over what was later found to be the brightest non-solar X-ray source, later designated Sco X-1. It also detected a uniform background glow which could not be resolved into individual sources. A follow-up campaign using X-ray detectors with better spatial resolution and optical telescopes identified Sco X-1 as an interacting binary with a compact (neutron star) primary.

This success led to further suborbital rocket flights by a number of groups. More X-ray binaries were discovered, as well as X-ray emission from supernova remnants, the radio galaxies M87 and Cygnus-A, and the Coma cluster. Detectors were improved and Geiger counters were replaced by proportional counters, which provided information about energy spectra of the sources. A constant challenge was determining precise positions of sources as only collimators were available.

The first X-ray astronomy satellite, Uhuru, was developed by Giacconi's team and launched by NASA in 1970. In its first day it exceeded the combined observation time of all previous X-ray astronomy experiments. Uhuru performed an all-sky survey using collimated proportional counters and detected over 300 individual sources. Among the discoveries from Uhuru were pulsations from X-ray binaries and extended X-ray emission from clusters of galaxies. The 1970s saw a succession of increasingly sophisticated satellites with X-ray detectors. Among them were Copernicus, Ariel-V (from the UK), ANS (from the Netherlands), OSO-8, and HEAO-1. These missions established further classes of astronomical objects as X-ray sources, observed more types of time variability from X-ray binaries, and detected iron emission lines.

   The next revolution in X-ray astronomy was wrought by the Einstein Observatory, launched in 1979 and named in honour of the centenary of his birth. X-ray focusing optics had been flown on Copernicus and as part of the solar astronomy experiment on Skylab but the Einstein Observatory provided the first X-ray images of many classes of astronomical objects. The combination of an X-ray telescope and an imaging proportional counter provided the sensitivity to observe large samples of stars, binaries, galaxies, clusters of galaxies, and active galactic nuclei (AGN). In addition, the focusing optics allowed the use of physically small detectors such as solid-state and crystal spectrometers as well as a grating that dispersed the spectrum onto a microchannel plate detector. The Einstein Observatory was one of the first astronomy satellites to have a guest observer program. Another innovation was an automated data-reduction pipeline and a public archive. Although this pre-dated the Internet, and thus required actual travel to the archive, it was an important first step.

   The 1980s were a lean period for X-ray astronomy in the USA but progress continued in Europe and Japan. Exosat was launched by ESA in 1983 into a deep 90-hour-period orbit which allowed long, continuous observations of sources. The Japanese scientific space agency, ISAS, pursued a program of placing mainly large-area proportional and scintillation counters on a series of satellites: Hakucho, Tenma and Ginga. Among the discoveries in this decade were quasi-periodic oscillations (QPOs) in X-ray binaries, iron emission and absorption lines from AGN and iron emission lines from the Galactic center. The successor to the Einstein Observatory was ROSAT, a German–US–UK collaboration with X-ray and extreme ultraviolet (EUV) telescopes, which was launched in 1990. The first six months were spent performing an all-sky survey, generating a catalog of more than 150 000 objects, followed by another eight years of targeted observations as part of a guest observer program, accumulating another catalogue of 100 000 serendipitous sources. ROSAT was able to image over a two-degree FOV providing good observations of large supernova remnants, clusters of galaxies, structure in the interstellar medium, and comets.

   The next big technological advance was the use of X-ray-sensitive charge-coupled devices (CCDs), which provide better imaging and spectroscopic properties than imaging proportional counters or scintillators. ASCA, launched in 1993 and a collaboration between ISAS and NASA, was the first X-ray astronomy satellite carrying CCDs. ASCA's other innovation was light-weight optics which provided a much larger area than those used for ROSAT although with poorer spatial resolution. The light-weight optics were placed in an extendable structure giving a long enough focal length to take the ASCA bandpass up to 10 keV, thus including the important 6–7 keV iron line region. This improvement enabled the detection and study of relativistically broadened iron lines in the accretion disks around black holes. Another major discovery made possible by the broad energy bandpass was that of non-thermal X-ray emission in supernova remnants.

While most X-ray satellites now use focusing optics, there is still a place for large-area proportional counters. RXTE, launched by NASA in 1996, was designed to collect many events from bright sources and to examine their variability down to microsecond timescales. An all-sky monitor tracked the brightest sources on day to year timescales. Among RXTE's discoveries were spin periods in low-mass X-ray binaries and kilohertz QPOs. RXTE's all-sky monitoring duties have now been taken over by the Japanese MAXI detector on the International Space Station.

In recent years, X-ray observations of gamma-ray bursts have become important. The Italian–Dutch satellite BeppoSAX, launched in 1996, had two sets of detectors. Proportional counters behind coded aperture masks were used to detect gamma-ray bursts and determine their approximate positions. These positions were used by ground controllers to point a set of narrow-field detectors at the source to observe the X-ray afterglow. This strategy was improved upon by the NASA satellite, Swift, launched in 2004, which autonomously points its X-ray and UV/optical telescopes at bursts detected by its wide-field coded aperture mask camera.

We are now in the era of great observatories with Chandra from NASA, XMM–Newton from ESA, and Suzaku from JAXA, all operational. Chandra's strength is its sub-arc second resolution telescope giving high-resolution images and, using gratings, spectra. XMM–Newton has the largest area of any focusing X-ray telescope while Suzaku covers a wider energy bandpass and has the lowest background. In the next few years, a number of other X-ray astronomy missions are planned. From NASA, NuSTAR will use focusing optics at energies above 10 keV for the first time and GEMS will measure X-ray polarization. A collaboration between Russia and Germany, Spectrum–Roentgen–Gamma, will perform a new, more sensitive, all-sky survey. Astro-H, the next JAXA mission, done in collaboration with NASA, will feature a microcalorimeter providing sensitive, non-dispersive, high-resolution spectroscopy. The international basis of X-ray astronomy is expanding as India, China, and Brazil all prepare their own satellites.

After its first half-century, observations in the X-ray waveband have become important in many topics in astronomy. Despite this, its techniques remain unfamiliar to many astronomers. One reason is that X-rays are almost always detected event by event instead of as a bolometric flux over a specific bandpass, as is common in other wavebands. However, the differences are not just in detection methods. While optical spectroscopy is concerned principally with line emission and absorption and radio spectroscopy with continuum emission, the processes that generate X-rays create significant continuum as well as line emission and both must be modelled correctly. Complicating this further, until recently, X-ray detectors did not have the spectral resolution needed to separate the lines from the continuum cleanly.

These and other issues have continued to make X-ray data analysis challenging. Following the launch of Chandra and XMM–Newton, the need for more systematic training of graduate students interested in X-ray astronomy became clear, leading to a series of X-ray "schools." These schools have been organised in the USA by the Chandra X-ray Center (CXC) and NASA Goddard Space Flight Center (GSFC), in Europe by the XMM–Newton Science Operation Centre (SOC), and in the developing world by the COSPAR Capacity-Building Workshop program. All of the authors of this handbook have lectured at these schools, in most cases multiple times, and the material collected here was developed through interactions with the students at the schools.

We have attempted to steer a middle course between pure theoretical exposition and explicitly detailing commands, albeit with excursions to both sides. We hope that this handbook proves to be a useful guide to beginning X-ray astronomers as well as experienced scientists who need to remember a conversion factor. To that end, we have arranged the text in three sections. The first three chapters cover the optics and detectors used on X-ray satellites. The next five chapters describe analysis issues, including data preparation, calibration, and modelling. Finally, the appendices contain a range of useful tables, including atomic data, conversion factors, and typical X-ray sources, amongst other information.

The editors thank Ilana Harrus for laying the initial groundwork for this book and Peter Wilmore for organizing the COSPAR workshops which provided much of our inspiration. We thank Adam Foster for creating a figure for Chapter 5, John Vallerga for help with Appendix 2 and Ken Ebisawa for help with Appendix 4. Nancy Brickhouse, Andrew Szentgyorgyi,

Harvey Tananbaum, Panayiotis Tzanavaris and Martin Zombeck provided helpful comments on the draft. The section on timing analysis leans heavily on X-ray astronomy talks given by Michael Nowak and Tod Strohmayer although they bear no responsibility for any errors introduced. Last, but not least, we thank our spouses for putting up with our distraction while finishing this book.

1

Optics

DANIEL A. SCHWARTZ

1.1 Introduction

It may be obvious why visible astronomy utilizes images, but it is illustrative to consider the value of focusing to X-ray astronomy. A list of advantages offered by the best possible two-dimensional angular resolution would include:

(i) Resolving sources with small angular separation and distinguishing different regions of the same source.
(ii) Using the image morphology to apply intuition in choosing specific models for quantitative fits to the data.
(iii) Using as a "collector" to gather photons. This is necessary because X-ray-source fluxes are so low that individual X-ray photons are detected; the weakest sources give less than one photon per day.
(iv) Using as a "concentrator," so that the photons from individual sources interact in such a small region of the detector that residual non-X-ray background counts are negligible.
(v) Measuring sources of interest and simultaneously determining the contaminating background using other regions of the detector.
(vi) Using with dispersive spectrometers such as transmission or reflection gratings to provide high spectral resolution.

The Earth's atmosphere completely absorbs cosmic X-rays. Consequently, X-ray observatories must be launched into space; so size, weight, and cost are always important constraints on the design. In practice this leads to a trade-off between the best possible angular resolution and the largest possible collecting area.

Realizing an X-ray telescope involves two key issues: reflection of X-rays, and formation of an image. We discuss each in turn.

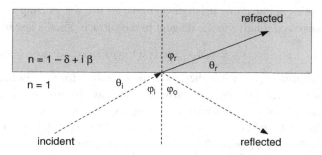

Fig. 1.1 An X-ray in the vacuum (unshaded lower region) impacts a medium (shaded upper region) at a grazing angle θ_i. When the real part of the index of refraction, $1-\delta$, is less than 1, the refracted angle θ_r is smaller than the grazing angle. Since $\cos \theta_r$ cannot be larger than 1, there is total external reflection for grazing angles less than $\arctan(1-\delta)$

1.2 X-ray reflection

X-rays incident on a material will, in general, penetrate and, if the material is thick enough, be absorbed. This is the basis of their utility in familiar applications such as medical diagnosis and luggage inspection. However, when impacting a surface at small grazing angles X-rays can reflect. An analogy is skipping stones on water instead of dropping them straight in. For X-rays the physical process is total external reflection, in which an individual photon collectively interacts with an ensemble of surface electrons. It is scattered coherently only in a special direction; namely, in accordance with the familiar law for all optical reflection, that the angle of incidence equals the angle of reflection, $\phi_i = \phi_o$.

1.2.1 Total external reflection

Figure 1.1 shows the interaction between an incident ray in the vacuum and a solid medium. The medium has an index of refraction $n = 1 - \delta + i\beta$, where δ and β are the optical constants (tabulated in e.g. Henke *et al.*, 1993). Both optical constants are very much smaller than unity in the X-ray regime. They are proportional to the square of the wavelength and the real and imaginary parts of the complex atomic scattering factor. If the medium is conductive then β is non-zero and the refracted ray will decay exponentially.

The incident and refracted angles are related by Snell's law:

$$\sin \phi_r = \sin \phi_i / (1 - \delta) \tag{1.1}$$

In X-ray optics applications it is conventional to use the grazing angle $\theta = \pi/2 - \phi$ measured relative to the tangent to the surface. Snell's law is then:

$$\cos\theta_r = \cos\theta_i/(1-\delta) \qquad (1.2)$$

Total reflection occurs when there is no real solution for θ_r in Equation 1.2. At X-ray wavelengths the optical constant δ is always positive so, since $\cos\theta_r$ cannot exceed unity, there is a critical grazing angle θ_c below which refraction is impossible and total external reflection occurs. The angle θ_c is given by $1 = \cos\theta_r = \cos\theta_c/(1-\delta)$, or $\cos\theta_c = (1-\delta)$. For small angles $1 - \delta = \cos\theta_c \approx 1 - \theta_c^2/2$ so that $\theta_c = \sqrt{2\delta}$.

At X-ray energies, E, which are not too near absorption edges of the reflecting material, the optical constant δ is given by:

$$\delta = r_e(hc/E)^2 N_e/(2\pi) \qquad (1.3)$$

where r_e is the classical electron radius, N_e the electron density, h Planck's constant and c the speed of light. Since N_e is proportional to the atomic number Z the critical grazing angle can be written:

$$\theta_c \propto \sqrt{Z}/E \qquad (1.4)$$

Thus, the critical angle decreases inversely proportionally to the energy. Higher-Z materials are desireable for reflecting surfaces since they have a larger critical angle at any energy and will reflect at higher energies at any fixed grazing angle.

The above discussion must be modified for energies just above atomic edges where absorption coefficients increase and reflectivity decreases. It is important to verify calculations experimentally in these energy ranges during calibration and bear in mind during data analysis that there may be additional systematic uncertainties.

The grazing angle will typically define the maximum angular size of the FOV. Current-generation X-ray telescopes typically have energy ranges from 0.1 to 10 keV using grazing angles of 1/2 to 1 degree. Materials such as gold, platinum, and iridium are good reflectors for these energy ranges. Lighter-weight materials such as beryllium, aluminum, and nickel were used in earlier telescopes designed for X-rays with energies primarily below 1 keV.

1.2.2 The Fresnel equations and reflectivity

After determining the conditions for reflection it remains to calculate the amplitude. Assuming a plane wave incident from vacuum onto an infinitely smooth surface then the components of $E_\parallel$, $H_\parallel$, $D_\perp$ and $B_\perp$ must be continuous across the interface (Jackson, 1988). Maxwell's equations then give the Fresnel equations:

$$r_p \equiv \frac{(n^2 \sin\theta - (n^2 - \cos^2\theta)^{1/2})}{(n^2 \sin\theta + (n^2 - \cos^2\theta)^{1/2})} \tag{1.5}$$

$$r_s \equiv \frac{(\sin\theta - (n^2 - \cos^2\theta)^{1/2})}{(\sin\theta + (n^2 - \cos^2\theta)^{1/2})} \tag{1.6}$$

where r_p is for the parallel component (i.e. in the plane of the incident and reflected photon) and r_s for the perpendicular component of the electric vector. The squared amplitudes of the complex numbers r_p and r_s are the reflectivities of each component. For unpolarized X-rays the desired reflectivity is just:

$$R = (|\, r_p\, |^2 + |\, r_s\, |^2)/2 \tag{1.7}$$

Since the squared amplitude of the refraction index, n^2, is very nearly unity in the X-ray range, polarization can be generally ignored and Equation 1.7 used for the reflectivity of a single surface. The high accuracy of this assumption is very convenient for designing a mirror, but has the unfortunate consequence that the measurement of X-ray polarization is very difficult.

1.2.3 X-ray reflection in practice

For the practical design, calibration, and analysis of data from X-ray telescopes, at least three other significant effects must be considered: scattering; interface of the mirror surface to the X-ray incident in vacuum; and preparation of the reflecting surface.

Surfaces are not infinitely smooth, as assumed in applying the boundary conditions for calculating the Fresnel equations. A bumpy surface will scatter incident X-rays. This scattering cannot be treated exactly because it is caused by deviations of a few Ångstrom on linear scales of a micrometer and less, which cannot be measured. Section 1.3.3 describes how to calculate scattering using a statistical description of the surface roughness. The key features to remember are that scattering is proportional to E^2, and that scattering in the plane defined by the surface normal and incident X-ray direction dominates over out-of-plane scattering by a factor $1/\sin\theta$.

The interface from a vacuum to a single, infinitely thick reflecting layer is thus an oversimplification. There typically is a substrate material which is formed and polished to the figure required for optical performance. For Chandra this is Zerodur, a special glass with an extremely small coefficient of thermal expansion. The high-Z metal which provides the high reflectivity is deposited on this substrate. However, materials such as gold or iridium require a binding layer, e.g. chromium, to hold the metal to the glass. Finally, there is an unwanted but unavoidable overcoat of molecular contaminants (e.g. carbon, oxides, . . .). At those energies where the reflectivity of the heavy metal is not near unity, i.e. high energies, large grazing angles, and above atomic edges, the X-ray penetrates the metal, so that the additional layers have a significant effect. One consequence is that reflectivity at some energies can be enhanced due to constructive interference from the different layers. This has been exploited to construct multi-layer mirrors for high-energy X-rays and for near-normal incidence mirrors at energies below 1 keV (see Section 1.3.5).

The index of refraction, and hence the Fresnel equations, assume the ideal density of a bulk material. However, the thin layers of reflecting metal, typically only a few hundred Ångstroms thick, do not necessarily have such a high density. Why not make a mirror out of pure gold? Aside from the cost of gold, it is a soft material with a relatively high coefficient of thermal expansion, which would not preserve a shape to the required specification under practical conditions. Also, since the weight that can be launched into space is a crucial limitation, a thick, high-density material would only allow very small mirrors. The density of the reflecting layer depends on the method of deposition. Sputtering generally results in a higher density than evaporative heating and deposition.

1.3 X-ray mirrors

Now that we understand how X-rays reflect, we can consider how to form surfaces which will bring an incident bundle of parallel rays to as close a point image as possible in the focal plane. X-ray astronomy studies sources such as supernova remnants, galaxies, and clusters of galaxies, which have sizes of many arc minutes, so focusing should be performed over a FOV of at least this extent. We consider first a simple parabola, and then the paraboloid and hyperboloid shapes, which have been the standard configuration for cosmic X-ray astronomy imaging.

1.3.1 Parabolas

Photons incident parallel to the axis of a parabola will all be reflected to converge at its focus. To verify this, consider that an incoming ray will hit the parabola at an angle $\alpha = \arctan dR/dz$ and be diverted through an angle 2α to follow a line defined by the equation $\tan 2\alpha = R/(z - f)$, where f is the distance from the origin to the focus, z is the coordinate along the parabola

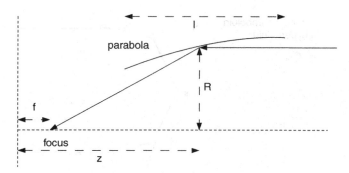

Fig. 1.2   Reflection from a parabolic surface. Rays parallel to the axis are focused
at the point f

axis, and R is the distance from the axis to the surface at the point of incidence
of the X-ray (see Figure 1.2). Using the trigonometric identity for $\tan 2\alpha$ in
terms of $\tan \alpha$ and substituting gives a differential equation in R whose solution
is the parabolic equation $R^2 = 4fz$. The focal length, $F = z - f$, can be set,
for example, by choosing that the mirror segment, of length l, have a specific
grazing angle α at a radius R: $F = R/2\alpha$.

It would be optimal if a single parabola could be used as an X-ray mirror;
however, parabolas have the characteristic that the image size increases linearly
with the off-axis angle. This implies that any source in the FOV will contribute
photons to the on-axis target. Nonetheless, there are two important applications
of parabolas to X-ray optics.

One use of the parabola is in the Kirkpatrick–Baez configuration. If a
parabola of length l is translated a distance h to form a curved plate, it will produce one-dimensional focusing of a point to a line. Another plate at right angles
can focus in the other dimension. This configuration is commonly used in syn-
chrotron radiation beamlines because of the simplicity of forming a surface in
only one dimension. For cosmic X-ray astronomy this is not used because of
the low fraction of the gross aperture which is available for collecting X-rays,
and because of the longer focal length required for a given aperture. A single
plate has an area $\alpha l h$, but this must be multiplied by the filling factor of the
crossed plates which is typically only of order 50%.

The second, and more important, use of a parabola is to form a paraboloid of
revolution about the optical axis, to be used in combination with a hyperboloid
in a Wolter I geometry as described in the next section. This gives an effective
aperture area of $2\pi R\alpha l$. Because α (in radians) is a small number, many sets
of mirrors (shells) with different radii R_i are nested within each other to give
a larger total area for intercepting the incident X-ray beam. Since the area of

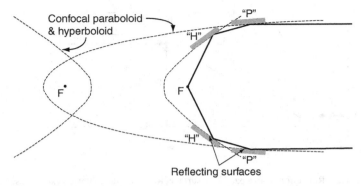

Fig. 1.3 Schematic of a Wolter type I X-ray mirror. X-rays incident from the right first hit a parabolic section and are redirected toward the parabola focus (left-hand "F"). If this is also one focus of the hyperbolic section which the X-rays hit next, then by the geometric definition of a hyperbola all the X-rays are directed to the other focus and form an image

the plate is $2\pi R l$, a ratio $\approx 1/\alpha$ more area must be precisely figured than is effectively available for collecting X-rays. For example, the Chandra effective area is about 1000 cm^2, similar to a 14-inch optical telescope. However, since the grazing angles are $25'$ to $55'$, almost 200 times more surface area had to be figured and polished.

1.3.2 Wolter's configurations

The classic papers of Wolter (1952a,b) proved that two reflections are needed to form an image over a finite FOV and considered the possible conic geometries which eliminated coma. The basic principle is that, in order to achieve perfect imaging, the optical path to the image must be identical for all rays incident on the telescope. Giacconi and Rossi (1960) first discussed the application of these geometries to X-ray astronomy.

Wolter derived three possible geometries. For types I and II the X-rays first reflect off the concave surface of a paraboloid, and then off the concave or convex surface of a hyperboloid, respectively. In type III the initial reflection is off the convex surface of the paraboloid followed by reflection from the concave surface of an ellipsoid.

Type I (Figure 1.3) is the geometry most useful in cosmic X-ray astronomy because it gives the largest aperture-to-focal length ratio. This ratio has been a key discriminant in maximizing the collecting area while staying within the relatively severe restrictions on length (and diameter) imposed by available space vehicles. For resolved sources, the shorter focal length concentrates a given spatial element of surface brightness onto a smaller detector area, hence gives a better signal-to-noise ratio against the non-X-ray-detector background. (This assumes the detector has sufficiently good linear resolution to sample the image properly.) There are structural advantages to the intersecting P and H surfaces. These include mounting, nesting, and vignetting considerations. For replicated mirrors, the P and H figures may be polished on a single mandrel and the pair formed as a single piece.

A grazing incidence telescope acts as a thin lens so, as the telescope tilts through small angles, the image of a point source near the axis remains fixed in space. This means a linear distance y in the focal plane can be converted to an angular distance $\theta = y/F$ on the sky. The thin-lens equation shows that the true optimal focal surface is a bowl shape, tangent to the flat plane perpendicular to the optical axis, and curved toward the mirror. In practice, a perfect on-axis image will not be formed due to optical imperfections in forming the mirror figure and the inevitable mechanical tolerances in aligning the various mirror shells. Consequently, it is generally advantageous to position a flat imaging detector slightly forward of the ideal on-axis focus in order to achieve better imaging performance over a finite off-axis FOV.

Recall that if α is the grazing angle of an on-axis X-ray hitting the parabola then it deflects at an angle 2α. The hyperbola is ideally formed with its surface at an angle 3α to the axis so that the grazing angle is again α. The reflected X-rays from an on-axis point source then converge to focus in a cone with a half-angle of 4α. The Chandra cone angles range from $3.417°$ to $1.805°$. For no more than a $0.1''$ contribution to the on-axis blur due to a focus error it was necessary to focus within $5\ \mu m/\tan(3.4°) = 85\ \mu m$. Figure 1.4 shows how this was achieved for Chandra on-orbit by off-setting the focal plane over a range of positions and finding the location giving the best auto-correlation of the resulting images of a quasar.

The effective area and image quality both decrease as a source moves off the optical axis. Figure 1.5 shows the Chandra effective area as a function of off-axis angle at two energies. Two effects contribute to these changes as a source moves off the optical axis: more photons are blocked by mutual shadowing of the nested shells; some photons will have improved reflectivity because they have a shallower grazing angle but this improvement is more than counterbalanced by other photons which reflect at larger angles. This latter effect also causes the relative effective area to fall off faster for higher energies.

Imaging cannot be perfect over a finite FOV, even for perfect optics and on the curved focal surface. Figure 1.6 shows the increase in the size of a point

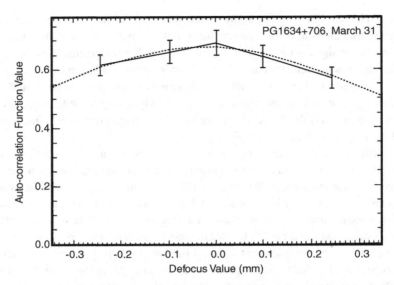

Fig. 1.4 Determining the focus of the Chandra mirror. By plotting the auto-correlation of a quasar image, in this case PG1634+706, vs. the defocus value, the best focus position is determined to within about ±50 μm, thus contributing less than 0.1″ to the total image blur

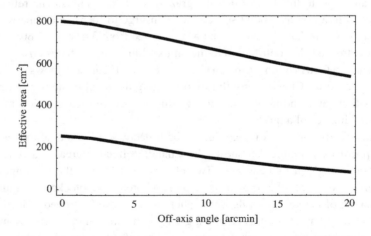

Fig. 1.5 The change in effective collecting area with off-axis angle, at 1.5 keV (top curve) and 6.4 keV (bottom curve). At higher energies, the relative area decrease is greater

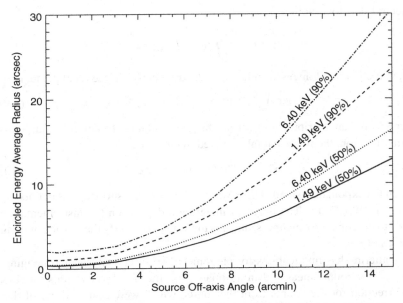

Fig. 1.6 The change in Chandra spatial resolution with off-axis angle, at 1.49 keV and 6.40 keV. The two curves for each energy show the effective angular size of a circle on the flat focal plane which contains 50% or 90% of the photons imaged from a point source. Source: Chandra Proposer's Observatory Guide

source as a function of off-axis angle, at two energies. At higher energies the imaging is always poorer due to scattering, as discussed in Section 1.3.3.

1.3.3 Scattering of X-rays

Scattering increases as the square of the X-ray energy, and typically dominates the image quality in the upper range of a telescope passband. For sufficiently smooth surfaces, scattering can be considered as diffraction, so that light of wavelength λ is scattered by irregularities which have spatial frequency f according to the usual grating equation: $f = \epsilon \sin \alpha / \lambda$, where ϵ is the scattering angle. Note that scattering is asymmetric. Backward scattering can be no more than $\epsilon = -\alpha$, as that would take the photon into the mirror surface. Forward scattering is unlimited.

The standard applications of scattering theory to X-ray astronomy (e.g. Aschenbach, 1985 or Zhao and Van Speybroeck, 2003) treat the irregularities in the surface height, h, at position x as random, characterized by a power spectral-density function:

$$2W_1(f) = \left| \int e^{i\,2\pi xf}\, h(x)\mathrm{d}x \right|^2 \qquad (1.8)$$

Then the scattered intensity, relative to the total power in the focal plane, is:

$$\psi(\epsilon) = 2W_1(f)8\pi(\sin\alpha)^4/f\lambda^4 \qquad (1.9)$$

For a Gaussian distribution of surface heights, and no correlation of roughness with direction, the relative total scattered intensity is:

$$1 - e^{-(4\pi\sigma\,\sin\alpha/\lambda)^2} \sim (4\pi\sigma\,\sin\alpha/\lambda)^2 \qquad (1.10)$$

when the exponent is small. The RMS roughness of the surface, σ is defined by $\sigma^2 = \int 2W_1(f)\,\mathrm{d}f$. The limits of integration depend on the instrumentation used to measure the roughness, and it is important to specify them when quoting values of σ.

Scattering has implications for photometry, spectral analysis, and imaging. To convert counts measured in an image to a flux in physical units requires a correction for the counts from the source which were scattered out of that image. When doing spectral analysis the correction becomes more critical, since the scattering fraction is a function of energy (see Equation 1.10 and Figure 1.6). Scattering causes a faint halo extended many arc minutes around a source. For most weaker sources this is not noticeable, but for a very bright source this surface brightness can overwhelm nearby faint features.

1.3.4 Contamination

The deleterious effect of dust on the surface of a grazing incidence X-ray mirror is enhanced by a factor $8/\tan\alpha \approx 8/\alpha \approx 800$ compared to the normal incidence case. This is because a dust grain of diameter d (assumed to be spherical on average) casts a shadow $d/\tan\alpha$ long. This is doubled since the grain can intercept an X-ray either before or after reflection, then doubled again because the X-ray must reflect off two surfaces. For grains of micron size and common composition (a mix of clay-like and plastics) it turns out that the absorption and scattering cross-sections are about equal resulting in the third factor of 2. A typical grazing angle is $\alpha = 0.01$ rad $= 34.4'$, which means that, to keep the degradation in throughput to no more than a few percent, less than a few times 10^{-5} of the area can be covered by dust. At such a level, which was achieved for Chandra, the energy dependence is negligible.

Contamination by molecular layers of organic material is unavoidable, and has a more complicated effect. Above the absorption edges of the constituents (typically C, N, O), the reflectivity is decreased due to absorption in the

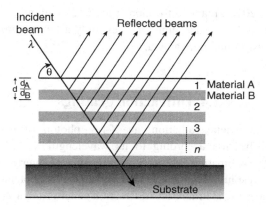

Fig. 1.7 Schematic of a multi-layer X-ray mirror. X-rays incident at angles θ larger than the critical grazing angle have a very small probability of reflection. However, a series of layers can be arranged so that the small reflection amplitudes can add coherently over a relatively narrow wavelength range, giving a significantly larger reflection

molecular material. However, above the absorption edges of the mirror materials the reflectivity is increased to some extent. In practice, minimization of contamination is a trade-off of performance requirements with costs. Most important for astrophysical analysis is that the contamination does not change after the mirror system is calibrated.

1.3.5 Multi-layer mirrors

Near-normal incidence reflection of low-energy X-rays is only of the order 10^{-4}, much too small for a practical telescope. This is because the X-rays penetrate the material until they are absorbed. Still, the amplitude of the reflected field vector is 10^{-2} to give a reflectivity of 10^{-4} so, if ≈ 100 layers were added coherently, then a significant reflection probability would be achieved. Such mirrors have been realized by alternating layers of a high-Z material, to provide high electron density for reflection, and a low-Z material, to provide a phase shift with minimal absorption (Figure 1.7).

Depending on the science objectives, it is either a bonus or a disadvantage that such mirrors typically have an energy bandpass of only 1% to 10%. Such mirrors have been used with great success to image the Sun in selected narrow energy bands (Golub, 2003).

Multi-layer surfaces have been studied intensively to make practical grazing incidence optics that can be used at energies of many tens of kiloelectron-volts (see e.g. Romaine *et al.*, 2004). They will be used on the NuSTAR mission

(Harrison *et al.*, 2010), planned for launch in 2012. Without multi-layers the critical grazing angles at such high energies, only a few arc minutes, would result in a uselessly small mirror area.

1.4 Diffraction gratings

Like all electromagnetic radiation, an X-ray photon of energy E keV will undergo diffraction according to its wavelength $\lambda = ch/E = 1.24 \times 10^{-7}/E$ cm. In combination with the phenomenon of interference, this can be used for high-resolution X-ray spectroscopy. In the focal plane of a thin lens the intensity of a classical Fraunhofer diffraction pattern from a slit of width a is $I_0(\sin \xi/\xi)^2$, where I_0 is the intensity on the optical axis, and $\xi = (a\pi/\lambda)\sin \alpha$, where α is the angle from the optical axis perpendicular to the length of the slit; i.e. in the dispersion direction. At a linear distance x in the dispersion direction, $x/F = \sin \alpha \approx \alpha$ for small angles, where F is the focal length. The intensity will be zero at angles such that $(a\pi/\lambda)\sin \alpha = m\pi$ for m a non-zero integer, and will have relative maxima at $(a\pi/\lambda)\sin \alpha = k\pi/2$ for k any non-zero odd integer.

In the presence of multiple slits, the diffracted beams from the individual slits will interfere, and modulate the single-slit pattern. If the slits are a distance d apart, then the path difference between adjacent slits for rays diffracting at an angle θ is $d \sin \theta$, and when this is an integer number of wavelengths, i.e. when $d \sin \theta = m\lambda$, all the rays are in phase at the focal plane and produce an intensity maximum (see Figure 1.8). The integer m, which may be positive, negative, or zero, is the diffraction order. Measuring the distance $x \approx F \sin \theta$ from the zero-order image measures the wavelength, or equivalently the energy, of the X-ray photon. If the path difference is λ/N for a total of N slits, then the phase difference, which is $2\pi/\lambda$ times the path difference, will sum to 2π for the N slits, and the intensity will go to zero. The ability of such a grating to distinguish adjacent wavelengths is measured by the resolving power $R \equiv \lambda/\delta\lambda$. The condition to resolve a line at $\lambda + \delta\lambda$ from the line at λ requires that it has an intensity maximum where the intensity from the line λ first vanishes. These two conditions are $d \sin(\theta + \delta\theta) = m(\lambda + \delta\lambda)$ and $d \sin(\theta + \delta\theta) = m\lambda + \lambda/N$, respectively. Equating the two right sides shows that $R = \lambda/\delta\lambda = mN$.

We have discussed slits of width a as if they were totally open, and spaced a distance d apart by intervening material which is totally opaque. Such a structure forms a transmission amplitude grating. For example, 0.5 μm of gold, or 14 μm of aluminum, will transmit less than 1% at 1 keV. Thus, a heavy

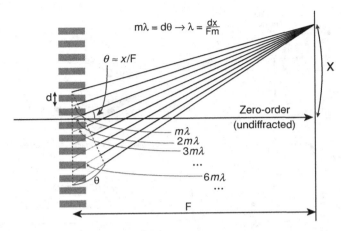

Fig. 1.8 A schematic (not to scale) of an X-ray-transmission diffraction grating. The grating spacing is d, typically ~ 0.2 μm, and the focal length F, typically $\sim$10 m

element such as gold is the material of choice. However, it is clear that the essential feature is that there are phase differences between the multiple slits. This principle is used both by phase gratings, where regularly spaced lines are produced by photolithography on a substrate, and by reflection gratings where grooves in a material of high atomic number can be shaped to optimize the reflected power into a particular order (Figure 1.9).

The above discussion has assumed perfect imaging by the optics, and implicitly that the detection process is perfect. In practice, the limiting factors are manufacturing imperfections in the optics and gratings as well as the spatial resolution of the readout detectors. In the grazing incidence configuration, which must be used for X-ray reflection, a transmission grating is comprised of a set of individual facets mounted to a fixture so that they intercept the converging rays exiting the mirror (Figure 1.10). To bring a given wavelength from any part of the grating to a common focus the individual grating facets are arranged on the Rowland torus, which is a circle in the plane of the optical axis and dispersion direction (Figure 1.11, top) and is rotated about the dispersion direction at the tangent point to the zero-order focus. This puts the long wavelengths progressively forward of the best imaging focus (Figure 1.11, bottom), degrading the image quality in the cross-dispersion direction.

X-ray astronomy almost always deals with continuum spectra, which may also have, or even be dominated by, emission or absorption lines. The energy of any photon is measured by its displacement x from the zero-order image. The X-ray flux at any energy is measured by summing the counts in rectangular

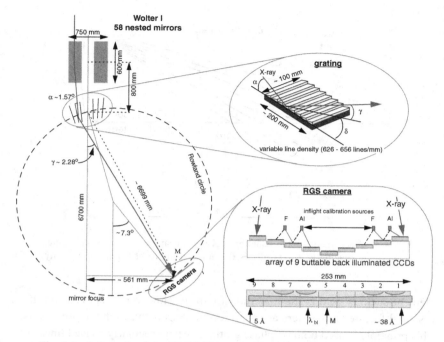

Fig. 1.9 Schematic of the XMM–Newton reflection grating. On the left is the arrangement of the mirrors (solid shading), with the gratings and RGS camera laid out on the Rowland circle. Top right shows X-rays reflecting from the blazed grating. Bottom right shows the RGS camera, a series of CCDs stepped to approximate the Rowland circle. From Brinkman *et al.* (1998)

regions of width equal to the energy resolution in the dispersion direction, and height equal to the spatial resolution in the cross-dispersion direction. There are two potential issues to beware of: order overlap and image overlap. From the equation for the intensity maxima, the second-order image of a photon of wavelength $\lambda/2$ (twice the energy) will end up superposed on the first-order image of wavelength λ. To avoid confusion, the detector itself must be able to measure and resolve photons with these two energies. Optical spectroscopy often positions an entrance slit at the focused image of the source, passing only a very restricted portion of the telescope FOV to the spectrometer. Since X-ray astronomy is generally photon-starved, the entire FOV of the telescope passes through the grating to perform slit-less spectroscopy. If the source is extended, the dispersed spectrum will comprise images of the source in each emission line. These images may overlap so different energies from different spatial regions can end up at the same detector position. For a relatively small and line-rich source the zero-order image may be used to fit distinct emission-line

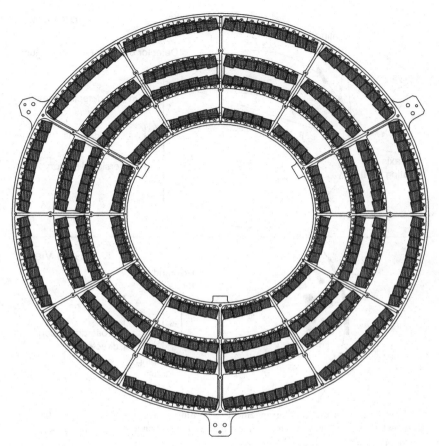

Fig. 1.10 The Chandra HETG plate holding the grating facets. *Source:* Chandra Proposers' Observatory Guide

images. If there are multiple sources in the field then the observation should be planned to ensure that their dispersed spectra do not overlap.

1.5 The future of X-ray optics

X-ray telescopes flown to date have had apertures equivalent to normal incidence mirrors of 10 to 30 inches in diameter. Scientific problems spanning astrophysics and cosmology require X-ray observations with collecting areas 30 to 1000 times larger, while retaining high-resolution imaging capability. New mirror technology is required to drastically lower the mass of X-ray reflectors,

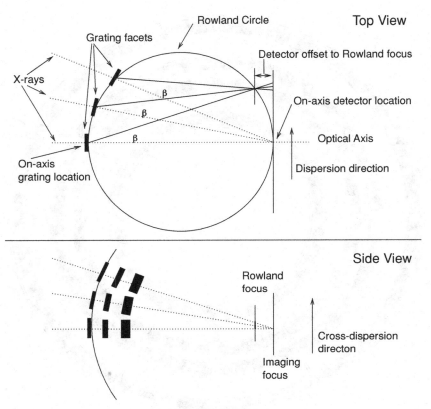

Fig. 1.11 Schematic of arrangement of grating facets and detectors on the Rowland circle for a transmission grating. Source: Chandra Proposers' Observatory Guide

in order to be able to launch the increased mirror sizes. The proposed International X-ray Observatory (IXO) is pioneering the technology of using very thin glass sheets, 0.4 mm thick, and slumping them on precision mandrels to form the reflecting figure. Because of the difficulties of mounting and aligning such mirror elements, the IXO team expects to achieve only 5″ half-power-diameter (HPD) angular resolution. To dramatically improve that resolution, the Generation-X vision mission conceives that it will be necessary to adjust the figure once on orbit, using thin piezo actuators deposited on the backs of the mirrors to impart local in-plane stresses. We hope for great progress in these areas as X-ray astronomy completes its second half-century.

2

Detectors

RICHARD J. EDGAR

2.1 X-ray detectors

Any experiment in X-ray astronomy must, of course, have a detector to register the X-rays, turn them into some kind of electronic signal, and, from there, into data that can be stored for later analysis.

The goal is to return as much information as possible about the X-rays incident on the detector. These data could include the energy of the X-ray photons, their arrival time, their number, the location where they were detected (in two dimensions if possible), and their polarization state. It is often the case that information about one of these properties must be traded for more precise information about another.

Nearly all X-ray detectors also make excellent detectors for all other types of ionizing radiation and, in some cases, even optical or infrared light. This means that the detector must also return enough information to discriminate between an X-ray photon and, for example, an energetic charged particle.

Since the environment of space is hostile and generally far from a repair shop, detectors must be made robust and reliable, able to function in vacuum, heat and cold, and, if the instrumentation involves high voltages, avoid shorting out at all reasonable pressures.

Essentially any material that is suitable for building an X-ray detector will interact strongly with X-rays somewhere in the very broad bandpass required in many instruments (for example, 70 eV to 10 keV in the case of the Chandra observatory). The X-ray region of the electromagnetic spectrum includes ionization edges for most materials. This means that the detector body itself will absorb some of the X-ray photons. Other items in the optical path will also interact with the photons in strongly energy-dependent ways. All of these effects must be calibrated, so that the observer will be able to convert detected counting rates into celestial photon fluxes or surface brightnesses.

This chapter introduces basic principles through a discussion of proportional counters, which have relatively simple design, then describes other types of detectors used in X-ray astronomy. We focus on those detectors important for archival and current observations but briefly mention some new developments. Charge-coupled devices are given a chapter of their own because of their importance.

2.2 Proportional counters

Proportional counters can be among the simplest X-ray detectors, and for this reason they were historically the workhorses of X-ray astronomy. A proportional counter can be flown without any focusing optic, but only a mechanical collimator, and still be useful in the measurement of the surface brightness of the sky, or the flux from bright sources. This type of detector can also be useful as a monitor during ground-based calibration of more complicated systems.

The first X-ray astronomy experiments were performed using Geiger counters but these were soon replaced by proportional counters. Proportional counters on suborbital rockets were used to observe many individual sources using collimators to restrict the field of view (FOV). This concept was refined for use on satellites starting with Uhuru (Giacconi *et al.*, 1971), then many others including the Exosat Medium Energy proportional counter (Turner *et al.*, 1981) and the Japanese satellite Ginga (Makino, 1987).

The group founded by William L. Kraushaar at the University of Wisconsin did a sky survey with a sounding rocket instrument of this type and, over the course of several years, mapped the diffuse, low-energy (≤ 2 keV) X-ray background in seven energy bands (McCammon *et al.*, 1983). The HEAO-1 satellite also conducted such a survey (Garmire *et al.*, 1992), as did SAS-3 (Marshall and Clark, 1984).

The currently active RXTE (Bradt *et al.*, 1991) satellite uses a technology of this type for the proportional counter array (PCA), which many scientists have found useful for timing X-ray pulsars and other variable sources.

Proportional counters have also been flown as detectors at the focal points of X-ray telescopes. Notable examples include the Einstein Observatory IPC (Giacconi *et al.*, 1979) and the ROSAT PSPC (Pfeffermann *et al.*, 1987).

2.2.1 Basic operation

At its simplest, a proportional counter is a grounded box of gas, with a high voltage wire through the middle, and a window of plastic or some other material which will admit most of the incident X-rays while keeping the gas in.

No thin window of plastic of any significant area can hold a differential pressure of the order of an atmosphere, so a fine mesh (not unlike window screen) and a system of struts must support the window. These structures are, of course, partially opaque to most X-rays.

At X-ray energies, the most probable interaction of a photon with matter is the photoelectric effect, in which the photon is absorbed by a single atom, and its energy goes into ejecting a single primary electron. This electron is usually from the K ($n = 1$) shell if there is sufficient energy available. Whatever energy remains over and above the K-shell electron's binding energy emerges as the kinetic energy of the resulting photoelectron. Note that the direction of travel of the primary electron is correlated with the polarization of the incident photon (see Section 2.2.7).

If the energy E of the original X-ray is large enough compared to the amount of energy it takes to produce an ion pair (an electron plus the ion it was removed from) in the gas, this leftover kinetic energy will be used by the primary photoelectron to ionize other gas atoms, producing a population of secondary electrons. If the energy needed to produce an ion pair is w, the number of primary and secondary electrons is given roughly by $N_e = E/w$.

Empirical values of w are around 28 eV for most commonly used gases and gas mixtures. Thus, for photon energies in the low-energy X-ray band, the number of electrons produced is rather modest ($N_e \sim 4$–400). This small number of electrons limits the energy resolution of the proportional counter. The process is not strictly Poisson (see Section 2.2.3). However, the variance of N_e is proportional to N_e, so the energy resolution $E/\Delta E \propto \sqrt{E}$.

This collection of primary and secondary electrons then begin to drift towards the high-voltage anode wires. As they move along the electric field, they accelerate. As soon as they have enough energy, they will ionize further gas atoms. In the volume within a few millimeters of the anode wires, the electric field increases quite rapidly, and an avalanche occurs, greatly amplifying the charge pulse. Depending on the gas mixture and the high-voltage setting, this amplification effect, called the gas gain, can be substantial: often two or three orders of magnitude.

The primary difference between a proportional counter and a Geiger counter is the voltage. In a Geiger counter, the anode voltage is high enough that the avalanche saturates, and the proportionality between incoming X-ray energy and pulse height (i.e. the charge collected in the pulse) is lost. In a proportional counter the voltage is tuned to maintain a linear relation between the input photon energy and the size of the charge pulse. This allows the proportional counter to be used as a crude spectrometer.

The resulting charge pulse is then shaped and amplified electronically. Some information about it, such as the time of arrival, the pulse height, and any position information that might be available, is recorded or telemetered to the ground.

2.2.2 Quantum efficiency

If it makes it through the window, an X-ray has a finite probability of making it through the gas volume as well. If the optical depths of the window and gas, respectively, are $\tau_w = \rho_w \mu_w(E) x_w$ and $\tau_g = \rho_g \mu_g(E) x_g$, then the probability that the X-ray interacts somewhere in the gas volume is given by:

$$P = e^{-\tau_w}(1 - e^{-\tau_g}) \tag{2.1}$$

Here ρ is the mass density of the material (in g cm^{-3}), $\mu(E)$ is the energy-dependent mass attenuation coefficient (in cm^2 g^{-1}), and x is the thickness of the material (in cm). The mass attenuation coefficient μ can be thought of as the interaction cross-section per gram of material, and is in fact just $\sigma(E)/m$ where m is the molecular weight of the material (in g).

The NIST Physical Reference Data Project (Hubbell and Seltzer, 2004) gives tables of the $\mu(E)$ functions for all elements likely to be used in experiments. These are updated from time to time, so for precise work it is important to get the latest versions, and to understand the experimental errors implicit in the data. The value of $\mu(E)$ is generally proportional to the inverse square or cube of the energy, with large jump discontinuities upwards at absorption edges, where the X-ray photon has just enough energy to ionize yet another electron energy level (Figure 2.1). Since X-rays predominantly interact with inner-shell electrons, and not with the valence electrons responsible for chemical reactions and bonds, the dependence of μ upon chemical structure is slight and quite subtle. Thus, to good accuracy, μ for a compound or a mixture is just the sum of the μ for the individual constituent atoms. However, the detailed structure of $\mu(E)$ near an absorption edge does depend on the chemistry, and can be easily resolved by the diffraction gratings in use in X-ray astronomy (Lee *et al.*, 2009).

The quantum efficiency (QE) is the probability, given that a photon strikes the front surface of a detector, that it is reported as an event. It comprises the probability P defined in Equation 2.1 modified by any other physical (or electronic) effect which can cause events to be lost. One example of such an effect is dead time (see Section 2.2.5).

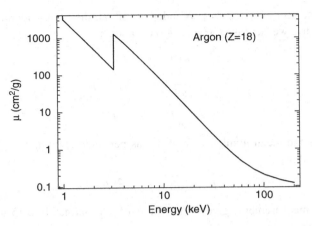

Fig. 2.1   Mass attenuation coefficient for argon (Hubbell and Seltzer, 2004)

It is sometimes useful to think of QE as having units of counts per photon. The detector turns X-ray photons into counts, with an efficiency given by the QE.

2.2.3 Energy resolution and the Fano factor

The energy resolution of a proportional counter is determined by the variances of two processes: the conversion of the incident X-ray into an electron cloud in the drift region and the charge amplification in the high-field avalanche region. The charge-amplification variance is proportional to the square of the gas gain. The variance on the initial conversion step is calculated as follows.

Recall that w is the average energy required to create an electron and an ion so the mean number of electrons produced by an incident X-ray of energy E is $N_e = E/w$. Naively, the variance on N_e would also be N_e, as for a Poisson process. In practice, however, the variance is much smaller for the following reason.

The total energy deposited must be E, so the ionization events are not independent. There are also many low-energy inelastic collisions (which excite but do not ionize the detector material).

Following Fano (1947), we now proceed to estimate the variance on N_e.

$$\sigma_e^2 = \text{var}(\hat{N}) = \langle (\hat{N} - E/w)^2 \rangle. \tag{2.2}$$

In this formula, $\hat{N} = \sum_{i=1}^{k} \hat{n}_i$ and $E = \sum_{i=1}^{k} \hat{e}_i$, where the random variables $\hat{n}_i$ and $\hat{e}_i$ are the number of ionizations and the amount of energy lost in each of the k interactions.

Since the set of $(\hat{n}_i - \hat{e}_i/w)$ are independent, all distributed identically, and of mean zero, we can take the sum of the squares:

$$\sigma_e^2 = k \langle (\hat{n}_i - \hat{e}_i/w)^2 \rangle$$
$$= \frac{E}{w\bar{n}} \langle (\hat{n}_i - \hat{e}_i/w)^2 \rangle$$
$$= \frac{E}{w} F = N_e \times F \tag{2.3}$$

where $\bar{n}$ is the mean number of ionizations per interaction and we introduce the Fano factor:

$$F = \langle (\hat{n}_i - \hat{e}_i/w)^2 \rangle / \bar{n} \tag{2.4}$$

Empirical measurements give values for F of approximately 0.15 for typical gases.

Combining the variance on N_e with that from charge amplification and converting from electrons to energy gives a one-sigma resolution of $w\sqrt{\sigma_{amp}^2 + \sigma_e^2}$, which in an ideal detector is about 15% at 6 keV.

2.2.4 Spectral response and escape

The spectral response of a proportional counter is described by the distribution of pulse-height values which can be produced for a given incident X-ray energy. This pulse-height distribution can be quite complex, even for a monochromatic X-ray source. The main peak, due to the photoelectric effect (Section 2.2.1) and sometimes referred to as the photopeak, has a width as derived in the previous section and an approximately Gaussian shape (Alkhazov, 1970). Additional features arise in the response as follows.

When an X-ray is absorbed by removing an inner-shell electron from an atom, the resulting ion can de-excite in several different ways. If another electron, ion, or atom collides with it, the excitation energy can go into kinetic energy, in a superelastic collision. If no collision is involved then the mostly likely mechanism is by emitting more electrons, a phenomenon called the Auger effect. Less likely, but still common, is for the ion to de-excite by emitting one or more photons. In this last case, it is particularly likely, when energetically possible, that an $n = 2 \rightarrow 1$ transition will be involved, producing a photon with an energy, $E_{2\rightarrow 1}$, somewhat less than the inner-shell ionization potential.

Since this photon does not have enough energy to ionize the same inner shell of the detector material, its range in the detector volume is larger than that of the original X-ray. There is a finite probability that it can escape from the

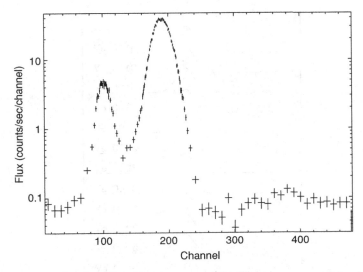

Fig. 2.2 Iron Kα photons incident on a flow proportional counter containing Ar and CH_4 gas. The peak near channel 195 is the primary peak at 6.4 keV. The Ar K escape peak can be seen near channel 100

detector volume, taking its energy with it. The resulting pulse of electrons will then be proportional, not to the original photon's energy E, but to $E - E_{2 \to 1}$.

For example, a gas commonly used in proportional counters is argon. The K-shell ($n = 1$) ionization potential for argon is 3206 eV. If photons with $E > 3206$ eV are incident on the counter, they can leave argon ions with a vacancy in the $1s$ state. If a $2p$ electron drops down to fill this vacancy, a photon of $E_{2 \to 1} = 2958$ eV can be emitted. If this photon escapes from the counter volume, the remaining energy is $E - 2958$ eV.

An example of a pulse-height spectrum is shown in Figure 2.2. The proportional counter was illuminated mostly by iron Kα photons, with energy 6404 eV. A peak around this energy (channel 195) is easily visible, but a significant minority of the events recorded have energy $6404 - 2958 = 3446$ eV (channel 100). This peak in the pulse-height spectrum is known as the escape peak.

The quantum efficiency is typically reduced by a few percent for energies just above the gas K-shell ionization potential because in this case the escape peak has so little energy that it is not registered by the detector so the incoming photon produces no signal.

Figure 2.3 shows proportional counter pulse-height spectra from a monochromator experiment conducted at BESSY in support of the calibration of

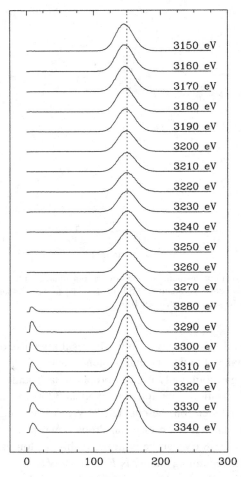

Fig. 2.3 Ar K edge scan with BESSY monochromator. These data were used to measure the flow proportional-counter gas opacity just below the edge. Listed energies are those requested; the true edge energy (indicated by the appearance of an escape peak and a jump in the strength of the main peak) is 3206 eV, thus calibrating the monochromator energy scale as approximately 75 eV too high. The dotted vertical line is merely to help discern relative peak positions

Chandra. The monochromator produces X-rays at a single energy which was stepped across the Ar K ionization edge (marked by a dashed line in the figure). The escape peak can be seen appearing at very low pulse heights.

In addition to the photopeak and escape peak the pulse-height distribution can also include fluorescence lines from elements in the gas or detector body as well as a tail or a shelf-like structure extending towards low pulse heights from the main peak. This last feature is due to subtle effects such as incomplete charge collection due to charge loss to the window, or to spatially variable gain.

The system gain (the conversion between mean pulse height and photon energy) can be non-linear, with measurable jumps or slope changes near edges of detector materials.

For these reasons, and the modest energy resolution, it is not, in general, possible to invert the pulse-height distribution to obtain directly an estimate of the input X-ray-photon energy spectrum. Instead, pulse-height distributions for monochromatic X-rays are collected into a calibration file called a response matrix file (RMF; Section 4.5.2). Logically, this can be thought of as a matrix with photon energy on one axis and pulse height on the other. A row of this matrix represents the distribution of probability over pulse-height channels, given that a photon is detected (e.g. Figure 2.2). Analysis techniques using the RMF are discussed in detail in Chapter 5.

2.2.5 Dead time and bright sources

Proportional counters can process events quite rapidly, but can only deal with one event at a time. There is thus a chance that an event will arrive while the previous event is still being processed digitally, or before the analog electronic signal dies away. The second event can then be recorded in a distorted form, or its charge can be confused with that of the first event. It can also be missed completely. The time that the detector is unable to register an event due to a previous event is called the dead time. The exposure time minus the dead time is generally referred to as the live time.

One way of monitoring the dead time due to the digital processing is to inject a pulser signal into the processing train, at the input to the electronic amplifier. A periodic signal with a pulse height in an otherwise unused portion of the spectrum (corresponding to an energy higher than the mirror cut-off, for example), can be counted both on input and in the output event stream. The fraction of injected pulser events which are detected on output gives an estimate of the live-time fraction.

Another technique for assessing problems is to monitor the time profile of events in the detector. If a second event arrives while the first transient is still in progress, it tends to lengthen the time taken by the event. Extended events can be rejected in hardware or flagged for careful examination during data analysis.

The dead-time fraction should also scale as the total event rate, whether all the events are reported or not. As discussed in the next section, various anti-coincidence techniques can be used to veto probable cosmic-ray events

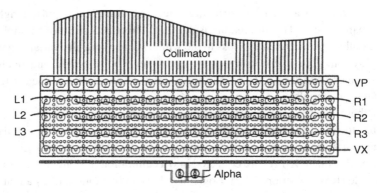

Fig. 2.4   Cross-section through a unit of the RXTE PCA. Each cell is surrounded by ground wires with an anode at its center. Anodes are linked together to give two signals from each layer (e.g. L1 and R1). The top and bottom layers as well as the side cells are linked together to provide additional background rejection

however such events still occupy the attention of the electronics for a finite period of time, rendering the detector blind until the processing finishes.

2.2.6 Position sensitivity and background rejection

So far the instruments discussed have been single-chamber, with only one detector volume. Such counters return no information about where on the window the X-ray struck, or where within the counter volume it underwent photoelectric conversion.

Somewhat more sophisticated designs can record information about the position of each event. This information is useful for forming images. If the detector is being used to read out a dispersive spectrometer, the position is used to reconstruct the energy of the input photon with great precision.

If there are several anode wires, each with its own amplifier, the ratio of the signals on the various anode wires gives information in one dimension about the position of the event.

The grounding of the circuit can also be used for position sensitivity. If the gas volume is subdivided by planes of grounded wires (again, each with an amplifier attached), the ground-plane signals can be used to obtain the location of the photoelectric event.

This subdivided counter design can also be used to improve background rejection. If the gas volume is divided into layers as in the RXTE PCA for instance (Figure 2.4), then X-rays will be stopped in a single layer while high-energy particles will travel through several layers, producing a signal in each. Detection of simultaneous signals can then be used (either in hardware or software) to veto events which are very likely to be particles and not X-rays.

Position sensitivity in two dimensions can be obtained by designing the counter with anode wires and two sets of cathode wires at right angles. A signal in an anode wire induces signals on nearby cathode wires which are used to determine the position. This method was used for the Einstein Observatory IPC and, greatly refined, for the ROSAT PSPC.

With position sensitivity in two dimensions, the edges of the counter volume can also be used as veto counters, thus providing five-sided veto capability. This design can greatly reduce the background due to high-energy charged particles, which are very abundant in near-Earth space environments. Their strong background rejection is one of the attractive features of proportional counters (and, as discussed below, microchannel plate detectors).

2.2.7 X-ray polarimetry

Gas-filled detectors can also be used to measure the polarization of X-ray signals. The direction of the primary photoelectron is correlated with that of the electric field of the incoming X-ray, with the number of events varying as $\cos^2 \phi$. An instrument design of this kind is planned for GEMS, a selected Small Explorer program satellite under development by NASA's GSFC (Swank *et al.*, 2009). The details of the instrument are described by Black *et al.* (2007).

The goal of such a detector is to track the trajectory of the primary photoelectron. This can be done using a detector geometry known as a "time projection chamber." A photon enters the gas-filled volume, ionizes a gas atom, and the photoelectron moves through the gas, ionizing as it goes. Each of these secondary electrons drifts toward an anode which is perpendicular to the entrance window. The anode is equipped with a multiplication stage: a set of small pores on two plates fixed at a large potential difference.

The one-dimensional readout gives the y-coordinate, parallel to the anode. The time of arrival gives the x-coordinate, perpendicular to the anode. In this way a track can be reconstructed (see Figure 2.5).

This arrangement produces a quantum efficiency comparable to that of a standard proportional counter and good sensitivity to the distribution of primary photoelectron angles.

2.3 Gas scintillation proportional counters

Another variation on the proportional counter is the gas scintillation proportional counter (GSPC). Examples of these have been the EXOSAT GSPC

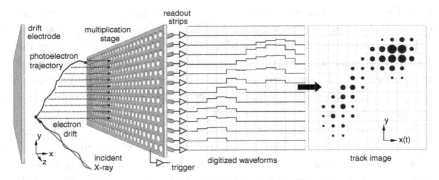

Fig. 2.5   A schematic representation of the GEMS instrument from Black *et al.* (2007). The drift time is much longer than the time to create the photoelectron track

(White and Peacock, 1988), the Tenma GSPC (Makino, 1987), the ASCA gas-imaging spectrometer (GIS) (Tanaka *et al.*, 1994), and the BeppoSAX HPGSPC (Manzo *et al.*, 1997).

A GSPC operates the same way as a proportional counter except that the electron avalanche and charge collection section is replaced by a scintillation region and photomultiplier tubes (PMTs). An incident X-ray ionizes an atom in the detector gas volume producing a primary photoelectron. This collides with other atoms, creating a secondary electron cloud which moves towards the bottom of the detector under the influence of the drift-region electric field. When the secondary electrons enter the high-field scintillation region they gain enough energy to excite the gas atoms (usually xenon), which then form diatomic molecules through collisions. The molecules de-excite by emission of UV photons, which pass through the exit window and are detected using the PMTs. The relative amounts of light detected in each tube can be used to estimate the position of the incident X-ray.

The main advantage of the GSPC over the standard proportional counter is that the variance of the scintillation step is less than that of the avalanche step. Consequently, the minimum resolution achievable by a GSPC is approximately 7% at 6 keV, twice as good as a proportional counter. The GSPC has now been superseded by CCDs for energies below 10 keV and CZT detectors (see below) for energies above 10 keV.

2.4   Scintillators

For energies greater than about 50 keV gas proportional counters become impractical because the counters must be built very deep or at very high pressure to provide a large enough column density of gas to stop an incident X-ray. One solution is to use a crystal which scintillates (i.e. emits optical light when absorbing an X-ray photon) then collect the resulting optical photons in PMTs. Combining two different crystals makes a phosphor sandwich or "phoswich" detector. A phoswich of NaI(Tl) on top of CsI(Na) was used for HEXTE on RXTE (Rothschild *et al.*, 1998) and PDS on BeppoSAX (Frontera *et al.*, 1997). In these detectors the NaI crystal functions as the X-ray detector with the CsI used for background rejection. Since the scintillation has different decay times in the two types of crystal it is possible to tell in which one the interaction occurred. An interaction in both crystals indicates an energetic particle (e.g. a cosmic ray) while a detection in the CsI and not the NaI is due to an X-ray coming in the side or bottom of the detector.

The higher-energy detector on Suzaku (HXD; Takahashi *et al.*, 2007) uses a phoswich of GSO (Ce-doped Gd_2SiO_5) crystals for X-ray detection and BGO ($Bi_4Ge_3O_{12}$) crystals for background rejection. The BGO completely surrounds the GSO except for a narrow opening filled with a collimator.

2.5 Microchannel plates

If very high spatial resolution is required but energy resolution is not so important then a good detector choice is the microchannel plate (MCP). This instrument consists of a plate composed of many small pores or tubes, side by side. Often they are inclined to the plate normal by an angle of a few degrees, so that incoming X-ray or UV radiation strikes one side of a pore. In many designs, pairs of plates are used in series. The plates are in an externally imposed high-voltage electric field, so that when a photoelectron is produced by an X-ray impacting near the front surface of the plate, an avalanche occurs in the pore and a pulse of electrons comes out the other end. These can be collected with anode wires, and their position determined with high accuracy, giving an imaging detector with good spatial resolution. Figure 2.6 shows a schematic diagram of the Chandra HRC.

The disadvantage of such a detector is that nearly all the spectral information is lost. The pulse-height distribution differs only slightly between low-energy (0.28 keV) and high-energy (6 keV) X-rays. There are, however, no consumables (such as gas for a proportional counter) or refrigeration requirements (as for CCDs and cryogenic detectors). The microchannel plate detector (the HRI) on ROSAT functioned for several years after the proportional counter (the PSPC) ran out of gas. The MCP's high spatial resolution and good QE for low-energy X-rays makes it a good choice of detector for long-wavelength X-ray and extreme UV (EUV) spectroscopy with gratings. The Chandra

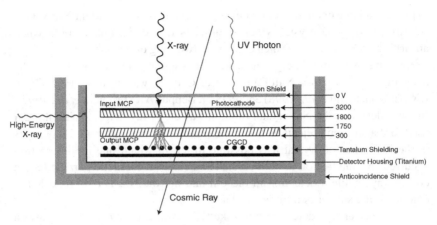

Fig. 2.6 A schematic illustration of the Chandra HRC

low-energy transmission grating (LETG) is usually used with the HRC-S as a readout device. The EUVE satellite also used MCP detectors in its spectrographs.

The quantum efficiency of an MCP can be greatly improved over bare glass by coating the front surface with a photocathode material; cesium iodide is a popular choice. While the QE is then higher at most energies, coating the plate introduces many more edges (of, for example, cesium and iodine). Calibration of the QE can be a challenge, especially as the MCP itself has very little energy resolution. Coating the detector also reduces the angular dependence of the QE, which arises from the angle between the detector normal and the pore axis, producing a variation with angle of how deeply into the pore the X-ray photon travels before ejecting a photoelectron from the walls.

There is a lifetime dose limit for an MCP, due to space-charge effects in the glass walls of the pores. When too much charge is extracted, the remaining positive space charge of the glass will make it difficult for electrons to be ejected. For example, the MCP camera that was used as an imaging detector during the ground calibration of the Chandra mirrors exhibited a QE deficit of tens of percent near the nominal aimpoint when the calibration was finished. To avoid this problem when operating in space, the spacecraft is often given a deliberate wobble so that the X-ray flux is distributed over many pores of the device.

As with proportional counters, guard detector volumes (often plastic scintillators) can be placed behind and on the sides of the instrument. This can produce an efficient background rejection, as cosmic-ray charged particles will often penetrate the detector and produce events in both the MCP and guard detectors.

Rapid digitization of the signal can allow for high-resolution timing analysis of the input X-ray signal (see Section 5.5). This type of instrument can be useful for study of objects such as X-ray pulsars, whose periods are often short compared to the ($\sim$ second) frame times of CCD detectors.

Microchannel plates are also excellent detectors of UV light. It is therefore necessary that detectors designed for X-rays screen out lower-energy photons. This is typically done with an aluminized plastic window near the MCP front surface. This UV-ion shield can also be used to control electrons scattered out of the MCP surface. Since MCP detectors do not need to be cooled, contamination of the window by outgassing from other parts of the satellite (and resulting QE degradation) is usually not a concern.

2.6 CZT detectors

The modern standard instrument for detecting X-rays at energies above 10 keV is the CZT (or CdZnTe) detector. It is used in the BAT on Swift (Barthelmy *et al.*, 2005) and will be the focal plane instrument on NuSTAR (Rana *et al.*, 2009). CZT is a semiconductor so it operates as an X-ray detector in the same way as the CCDs described in the next chapter. The difference with a CCD is that each pixel must be read out independently, since there is no transfer of charge along rows or columns. As in a CCD, an incident X-ray creates electron–hole pairs with the energy required to create a single pair being 1–2 eV. Energy resolutions of a few percent are achievable in the 10–100 keV energy range. The reason to use a CZT detector instead of a CCD at these energies is that CZT has a much higher cross-section than Si so the detector has a better QE.

2.7 Microcalorimeters

High-resolution X-ray spectroscopy is achieved on Chandra and XMM–Newton using gratings to disperse the spectrum onto CCDs or MCPs. While this has had some great successes, as a technique it suffers from two shortcomings. X-ray gratings have low efficiency which limits their use to relatively bright sources and, having no slit, they do not perform well on extended sources. A technology has been developed over the last couple of decades which solves both these problems. The quantum microcalorimeter measures the change in temperature of a small block of material when it absorbs an X-ray. This temperature change can be measured to great precision, providing an energy resolution of a few electron-volts at 6 keV. The difficulty with microcalorimeters is that the change in temperature that is being measured is a few millikelvins so the detector must be cooled down to the tens of millikelvin range, requiring a complex cryogenic system. A microcalorimeter (XRS; Kelly *et al.*, 2007), was flown on Suzaku and achieved a resolution of 7 eV during initial calibration. However, before any astronomical observations could be made, there was a catastrophic loss of the liquid-helium cryogen due to a spacecraft design error. X-ray microcalorimeters have been flown on suborbital experiments and are being used for laboratory astrophysics. The next attempt to fly one on a satellite will be on Astro-H, due for launch by JAXA in 2014.

3

Charge-coupled devices

CATHERINE E. GRANT

3.1 Introduction

Charge-coupled devices, or CCDs, were invented at Bell Laboratories, New Jersey, in 1969. The advantages of CCDs for optical astronomy over the previous technologies were quickly realized and the use of CCDs revolutionized astronomy in the 1980s due to their sensitivity and linear brightness response. CCD cameras are now the most common detector at optical observatories around the world and are the sensing element in nearly all commercial digital cameras.

It was also recognized early on that CCDs were sensitive to X-ray radiation as well as optical light, although optimizing the technology for X-ray use took longer. The first suborbital rocket flight equipped with an X-ray CCD camera was launched in 1987 to observe SN 1987A. Japan's ASCA (Tanaka *et al.*, 1994), launched in 1993, was the first satellite with an X-ray CCD camera. Since that time, CCDs have become ubiquitous in X-ray astronomy and are part of the focal-plane instrumentation in almost all recent past, current, and planned missions. The CCD detectors on the largest of the currently operating missions are ACIS (Garmire *et al.*, 2003) on Chandra, EPIC (Turner *et al.*, 2001; Strüder *et al.*, 2001) and RGS (den Herder *et al.*, 2001) on XMM–Newton and XIS (Koyama *et al.*, 2007) on Suzaku. Other CCD detectors on currently operating missions are XRT (Burrows *et al.*, 2000) on Swift and SSC on MAXI (Matsuoka *et al.*, 2009).

Understanding the basic principles of CCD operation and some of the resultant features encountered in data analysis is important for accurate interpretation. This chapter begins with an outline of the physical principles behind CCD operation in the X-ray regime, then describes how to quantify CCD performance to calibrate data. Next, some features of X-ray CCDs that observers need to consider when planning X-ray observations and analyzing X-ray data are discussed. The final section describes some improvements that may be

39

incorporated in future missions. While an attempt will be made to be as generic as possible when describing both the hardware and the language used in data analysis, the author is most familiar with ACIS on Chandra, so most examples will be drawn from that instrument.

3.2 Basic principles and operation

A CCD is an array of linked capacitors. Photons interact in a semiconductor substrate and are converted into electron-hole pairs. An applied electric field is used to collect the charge carriers (usually electrons) and store them in pixels. Each pixel is coupled to the next and can transfer stored charge to its neighbor. The stored charge is transfered, pixel to pixel, from the interaction point to a readout amplifier. At the readout amplifier, the charge is sensed and digitized. The remainder of this section will further describe each step in this process, but will not fully delve into semiconductor detector physics. Interested readers are encouraged to explore the many excellent textbooks on the subject for further details (Janesick, 2001; Lutz, 2007; Sze, 2002).

3.2.1 Photoelectric absorption

The first step in the process is the photoelectric absorption of an X-ray in a semiconductor substrate, usually silicon. As a reminder, solids can be classified by their conductivity into three categories: insulators, semiconductors, and conductors. The conductivity of insulators is very low, while for conductors it is high. Semiconductors are an intermediate case, and the conductivity is sensitive to many factors such as temperature, illumination, and the presence of small amounts of impurity atoms.

As an X-ray travels through the semiconductor substrate, it can be photoelectrically absorbed with a probability that is constant per unit length. The inverse of that probability is the attenuation length or mean absorption depth. Figure 3.1 shows the attenuation length in microns of an X-ray photon in silicon as a function of energy (or wavelength). The attenuation length is the average distance a photon of a particular energy will penetrate before interacting. The depth of the active region of a typical X-ray CCD is also shown in Figure 3.1, which then defines the energy range for best X-ray sensitivity, from roughly a few hundred eV to 10 keV. Photons with energies below a few hundred eV will interact close to the surface and may be incompletely detected or may be absorbed in any structures on the surface and not detected at all. Photons with high energies are likely to pass through the active region without interacting at all.

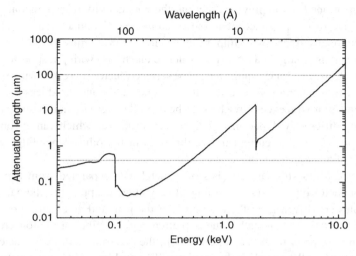

Fig. 3.1 The attenuation length or mean absorption depth of an X-ray photon in silicon as a function of the photon energy and wavelength. The horizontal dotted lines indicate typical active regions in an X-ray CCD which correspond to an energy range of about a few hundred eV to 10 keV. The best X-ray sensitivity will be in this range

The photoelectric interaction of an X-ray generates electron-hole pairs. On average, the number of liberated electrons is linearly related to the energy of the incident X-ray:

$$N_e = E_X/w \tag{3.1}$$

where N_e is the number of electrons, E_X is the energy of the X-ray photon and w is the ionization energy required to create an electron-hole pair. For silicon at typical X-ray CCD operating temperatures, w is roughly 3.7 eV per electron, so a single X-ray will generate tens to thousands of electrons. Once created, this charge cloud will diffuse and eventually recombine, or can drift under the influence of an applied electric field.

3.2.2 Charge collection

Once the X-ray photon has created a charge cloud, it must be collected and stored. The basic element of a CCD is a capacitor, either a metal-oxide-semiconductor (MOS) structure or a *p–n* junction. Chandra's ACIS utilizes MOS CCDs while the XMM–Newton EPIC detectors use both types. In a general sense the charge collection and storage of the two technologies is similar but the physical structures are quite different.

As mentioned in the previous section, the conductivity of a semiconductor is sensitive to the presence of impurity atoms. Semiconductor material can be intentionally doped with impurities to alter its properties. By replacing a small fraction of silicon atoms (four valence electrons) with phosphorus (five valence electrons), for example, an n-type semiconductor is created with excess free electrons. Doping with boron (three valence electrons) produces a p-type semiconductor with excess free holes. The overall material remains uncharged. Layering differently doped materials creates structures which can encourage or discourage electric current in one direction or the other depending on the polarity of any applied voltage.

A basic MOS structure consists of a metal gate separated from a p-type semiconductor by an insulator. If a positive voltage is applied to the gate, the free holes are driven from the region under the gate and an electric potential well is created. This is called the depletion region. When a photon creates electron-hole pairs in the depletion region, the electrons will be collected in the potential well just below the insulator. The problem with this structure is that charge trapping can occur at the semiconductor–insulator interface which harms performance as these electrons are not easily transferred to the readout. To prevent this, an additional layer of n-type semiconductor is placed between the p-type and the oxide. This pushes the deepest part of the potential well away from the oxide, so the electrons are collected in a buried channel. Figure 3.2 is a schematic of the electrostatic potential distribution in a buried-channel CCD. (An MOS structure could also be made with an n-type substrate and holes as the charge carrier. Historically, X-ray CCDs have been p-type, but the advantages of n-type, such as increased radiation hardness and improved high-energy quantum efficiency, may lead to more n-type devices in the future.)

As stated previously, the physical structures in a CCD based on a p–n junction are different but the general principles are the same. From an observer's viewpoint, the most important difference is that the material is fully depleted. A fully depleted p–n CCD can have a much thicker active region than a standard MOS CCD, yielding superior high-energy detection efficiency. Additionally, they can be back-illuminated without any thinning, yielding better low-energy detection efficiency (back-illumination is further discussed in Section 3.3.3). Conversely, rejecting charged-particle events can be less efficient for a fully depleted CCD, as discussed in Section 3.2.5.

X-rays that interact in the depletion region create free electrons that are collected in the potential well under the metal gates. The most common type of CCD is a three-phase CCD with three gates defining one dimension of each pixel (see Figure 3.3). The gates run the length of the CCD. The central gate is set to a positive voltage, while at least one of the neighboring gates is set

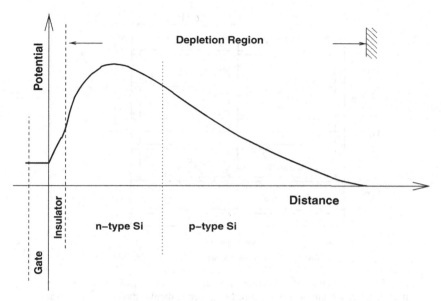

Fig. 3.2 A schematic of the electrostatic potential distribution in a buried-channel CCD (not to scale). X-rays that interact in the depletion region create electrons that are collected in the potential well

to a lower voltage. The potential well under the central gate is deeper than its neighbors, so the electrons will be collected there. The pixel size along this direction is defined by these gates and can be tens to hundreds of microns. In the other direction the pixel boundaries are defined by channel stops, a physical barrier of insulator and highly doped semiconductor below and perpendicular to the gates. A two-by-two-pixel schematic is shown in Figure 3.3.

3.2.3 Charge transfer

Charge is collected under a single gate for a period of time called the frame time. To transfer the charge, a higher voltage, which produces a deeper potential well, is applied to adjoining gates sequentially. A small overlap in time allows for the charge to move from one gate to the next and one pixel to the next. The name, charge-coupled device, refers to this brief joining of the adjacent potential wells to transfer charge. A popular metaphor for the CCD transfer process is a bucket brigade, in which a line of people transfer water from one end to the other by pouring water from one bucket to the next in sequence. In this fashion, charge can be transported over many hundreds of pixels in a single direction. As

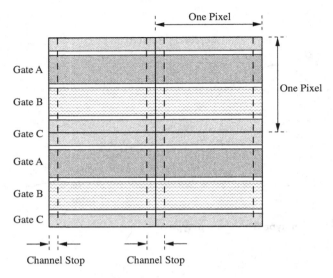

Fig. 3.3 Schematic of the structures that define the dimensions of a pixel. In the transfer direction are the three gates. In the perpendicular direction are channel stops beneath the gates. Source: Science Instrument Calibration Report for the AXAF CCD Imaging Spectrometer (ACIS)

the gates extend over the entire CCD, all the columns are transferring charge simultaneously in parallel, thus the step is often referred to as the parallel transfer. The end of each column can either terminate in a readout amplifier (as in the EPIC-pn detector on XMM–Newton) or an additional perpendicular array (the serial register) that transfers the charge from each column to a single amplifier.

Generally, CCDs for X-ray astronomy use frame transfer. The data are accumulated in the image section of the array which is open to X-rays. After the frame time is over, the charge in the entire image section is transfered quickly row-by-row in parallel to the framestore section. The framestore array is shielded to prevent X-ray interactions during the readout. Then each row in the framestore is transfered to the serial readout register, and each pixel is shifted serially to one or more output amplifiers and read out. A schematic representation of a frame-transfer CCD is shown in Figure 3.4.

3.2.4 CCD focal planes

A typical X-ray CCD in use today is at most a few centimeters square. Depending on the telescope plate scale, a single CCD may be insufficient to cover the

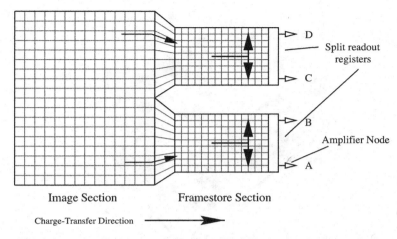

Fig. 3.4 Schematic representation of an ACIS CCD showing the frame-transfer architecture and the four readout nodes. The schematic is rotated such that columns are in the horizontal direction and rows in the vertical. Source: Science Instrument Calibration Report for the AXAF CCD Imaging Spectrometer (ACIS)

desired FOV. This is especially important when the CCDs are being used as a readout detector for a dispersive spectrometer which may spread the spectrum out over 10 cm or more. In either case, multiple CCDs can be tiled to increase the physical size of the focal plane. Figure 3.4 shows how the readout electronics are concentrated on a single side of the CCD. The remaining three sides are much less encumbered allowing two (or more) CCDs to sit closely together.

For example, Chandra ACIS consists of the imaging and spectroscopic arrays (Figure 3.5). The imaging array (ACIS-I) uses four CCDs to make up a 16 arcmin square array. These CCDs are tilted in a bowl shape to best follow the telescope focal surface (see Section 1.3.2). The spectroscopic array (ACIS-S) of six CCDs in a line is used as a readout detector for the HETG (or, less commonly, the LETG). These CCDs are tilted to approximate the Rowland Circle of the gratings.

Other multi-CCD focal planes are the XMM–Newton EPIC and RGS, and the MAXI SSC, which each use different tiling schemes. The Suzaku XIS and Swift XRT both have a single CCD in each focal plane. The advantage of using multiple CCDs is primarily the larger physical size although it does provide some protection against losing the entire focal plane due to a single malfunctioning CCD. On the other hand, the effective exposure will be greatly reduced (sometimes to zero) in the gaps between CCDs.

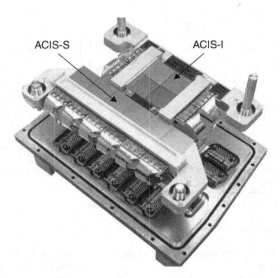

Fig. 3.5   A photograph of the ACIS focal plane. ACIS consists of two CCD arrays, the 2 × 2 imaging array and the 6 × 1 spectroscopic array. Each CCD is approximately 25 mm on a side. Image credit NASA/CXC/SAO

3.2.5 CCD operation

X-ray CCDs are generally operated in photon-counting mode where the position and the energy of every X-ray photon can be determined individually. This is different from CCDs in optical applications in which each photon produces too little signal to be independently sensed and longer frame times are used to accumulate many thousands of photoelectrons. Accurate spectroscopy in photon-counting mode requires that no more than one photon interacts in a pixel in a frame time. The minimum frame time of a CCD is limited by the readout rate, which in turn is limited by a tradeoff between readout rate and noise. For example, the ACIS instrument on Chandra has 1024 by 1024 pixels and four readout nodes per CCD with a 100 kHz readout (10 µsec per pixel), which limits the minimum frame time to about three seconds. This frame time can be reduced by reading out only a subarray of the full CCD and discarding the remaining pixels. By sacrificing the information from one spatial dimension, the time resolution can be improved further with continuous parallel clocking. For ACIS this implies the 3.2-second frame time can be reduced to as low as 0.4 seconds by reading out an eighth of the CCD or 128 rows, or to 2 milliseconds with continuous clocking.

Spacecraft are limited in the amount of telemetry they can transfer to ground stations on the Earth. The output rate of data from an X-ray CCD is very high, of order 10 Mbits/sec/CCD, and exceeds any reasonable telemetry resources. Because of the short frame time and typically low source count rates, most pixels in a CCD frame are empty. Rather than waste telemetry bandwidth sending empty pixels to the ground, the raw CCD frames are processed on-board to select candidate X-ray events. This event selection process determines and removes the CCD bias level, the signal in each pixel in the absence of external radiation, and compares the pixel pulse heights to a predetermined event threshold. If the pixel exceeds this threshold and is higher than its neighboring pixels, it is recognized as an event. For each event, the position, time and pulse height of the central pixel, and the pulse heights of the surrounding pixels, or event island, are recorded. On ACIS, this is a 3×3 pixel region. Each event is assigned a grade or pattern which characterizes the morphology of the pulse heights in the event island (see also Section 4.1.2 and Figure 4.2).

Depending on the relative sizes of the charge cloud and the CCD pixels, the charge from an X-ray event may be split over multiple pixels. The size in pixels of a valid X-ray event will depend on the characteristics of the CCD, such as front- versus back-illuminated and physical pixel size, and the energy and precise position of the original X-ray photon. In order to recover the original photon energy, the charge needs to be summed over all the split pixels. In doing so, the noise in each summed pixel is included as well. X-ray events that are split over multiple pixels suffer from more noise and incomplete charge collection than single events, so have poorer spectral resolution. Each mission determines a set of good event grades that best balances quantity of X-ray events with the quality of their spectral resolution.

This event morphology can also be used to discriminate between legitimate X-ray events and events associated with the charged-particle background. Figure 3.6 is a raw ACIS CCD frame showing the different appearance of X-ray events and charged-particle events, such as cosmic rays impacting the detector. The particles produce characteristic blobs and streaks while the X-ray events are small and point-like. The physical reason for this is that an X-ray photon interacts once and the photoelectrons travel very short distances before collection. A charged particle, on the other hand, produces electron-hole pairs along its entire path, including in any undepleted material. The charge deposited in the undepleted region diffuses to much larger sizes before encountering and partially collecting in the potential well and thus produces the large diffuse blobs. Therefore, the smaller and simpler event shapes, single or singly split pixels, are more likely to be X-rays, while larger shapes are more likely to be due to cosmic rays. Filtering event lists by event shape or amplitude on-board the spacecraft is often used to reduce telemetry by removing suspected charged-particle events. In this case the observer does not receive the full event

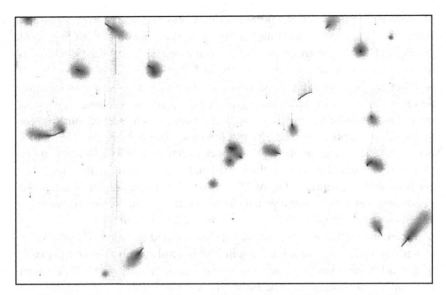

Fig. 3.6   A portion of a single raw CCD frame from ACIS taken on-orbit. The image scaling is logarithmic. The large blobs and streaks are due to charged particles while the small dots are individual X-ray photons. This image is from a front-illuminated CCD. The charged-particle morphology on a thinned or fully depleted device would look different

list as recorded by the instrument, but may receive ancillary data that record characteristics of the rejected events. Filtering by event shape in data analysis can reduce the background further, at the cost of eliminating some real X-ray events. Devices that are thinned or fully depleted have different responses to charged particles and, in particular, lack the largest diffuse structures, making discrimination between background and X-ray events somewhat more difficult.

3.3 Performance

Using this basic understanding of the physical principles and practices of X-ray CCD operation, the characteristics of CCDs that are important in planning, executing, and analyzing an observation can be discussed.

3.3.1 Quantum efficiency

The CCD detection efficiency or quantum efficiency is the product of the transmission through the gates, insulators, and channel stops, which act as dead

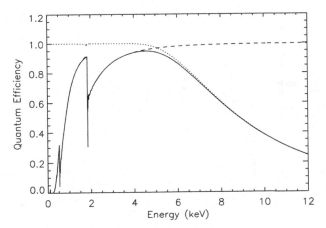

Fig. 3.7 The quantum efficiency of a CCD (solid line) is the product of the transmission probability through various dead layers (dashed line) and probability of absorption in the depletion region (dotted line)

layers and absorb incident X-rays before they can reach the photo-sensitive region, and the absorption in the depletion region where the X-ray can be detected. The transmission, T, through all the dead layers is the product of the transmission through each dead layer,

$$T = \Pi_i e^{-\mu_i t_i} \tag{3.2}$$

where μ_i and t_i are the linear absorption coefficient and the thickness of dead layer i. The absorption in the depleted region is then:

$$A = 1 - e^{-\mu_{Si} d} \tag{3.3}$$

where μ_{Si} is the absorption coefficient for silicon and d is the thickness of the depletion region. The resulting quantum efficiency is the product of the transmission and the absorption:

$$QE = (1 - e^{-\mu_{Si} d}) \, \Pi_i e^{-\mu_i t_i} \tag{3.4}$$

Figure 3.7 shows the transmission and absorption curves for a typical CCD, along with the resulting quantum efficiency. Quantum efficiency at low energies is dependent on the thickness of the dead layers, while at high energies it is dependent on the depletion depth.

As is obvious from their frequent use in ground-based astronomy, CCDs are highly sensitive to optical photons. In an X-ray detector, optical light is a source of noise and causes pulse-height calibration issues. For this reason, X-ray CCDs usually have an optical blocking filter. These are constructed

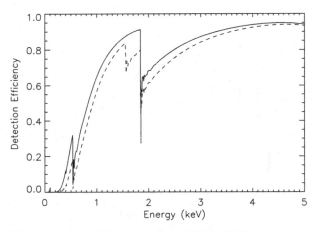

Fig. 3.8   The detection efficiency of a bare CCD (solid line) and the detection efficiency after including the loss from transmission through the optical blocking filter (dashed line)

of thin layers of aluminum and plastic a few hundred nanometers thick. The optical blocking filter also absorbs low-energy X-rays and acts as an additional dead layer. Figure 3.8 shows the detection efficiency of a bare CCD and the detection efficiency after including absorption by an optical blocking filter. At low energies below 500 eV, the filter reduces efficiency by more than 50%.

3.3.2 Energy scale and spectral resolution

As stated earlier, the interaction of a single X-ray photon with a silicon atom produces free electrons. The average number of electrons is a linear function of energy with w in Equation 3.1 roughly equal to 3.7 eV per electron. After the electrons are transfered to the readout amplifier, sensed, and digitized, the resulting value is called the pulse height (for historical reasons) in units of ADU (analog-to-digital unit) or DN (data number). The pulse height is a linear function of N_e, the number of electrons, and therefore also a linear function of the input X-ray energy. For X-ray events with charge split over multiple pixels, the event pulse height is summed over the relevant pixels. The scaling between energy and pulse height is the instrument gain.

The variance on the charge liberated is smaller than Poisson since the production of each free electron is not independent. The variance is:

$$\sigma_e^2 = F \times N_e \qquad (3.5)$$

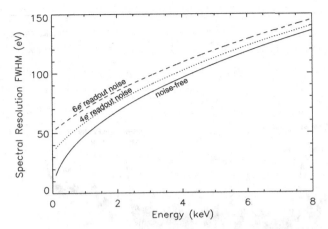

Fig. 3.9 The theoretical spectral resolution of an X-ray CCD for RMS readout noise of 6 e⁻, 4 e⁻ and for a noise-free detector. Modern X-ray CCDs typically have readout noise less than 5 e⁻

where F is the Fano factor introduced in Section 2.2.3 and is roughly 0.12 for silicon. The spectral resolution, or full width at half-maximum (FWHM), of the detector then depends on the readout noise and the physics of secondary ionization:

$$\text{FWHM(eV)} = 2.35w\sqrt{\sigma_e^2 + \sigma_{\text{read}}^2} \tag{3.6}$$

The w term converts the resolution from electrons to electron-volts. To achieve the best spectral resolution, a CCD needs to have good charge collection and transfer efficiencies at low signal levels combined with low readout noise. It also requires a high readout rate so that all pixels have at most one event per frame time. Figure 3.9 shows the spectral resolution as a function of energy for a noise-free detector and at different noise levels. Modern X-ray CCDs typically have readout noise less than five electrons (5e⁻).

In addition to the primary photopeak, the instrument response produces a number of other spectral features that can be seen in Figure 3.10. In the figure, the X-ray source produces several spectral lines of manganese: the Kα and Kβ lines and an unresolved complex of L-lines. The remaining structures are due to the instrument. Two of these features, the escape and fluorescence peaks, were introduced in Section 2.2.4. Every incident photon with an energy above the silicon K-edge (1.84 keV) has a small probability of producing a Si K fluorescence photon (1.74 keV). The remaining energy from the original X-ray event goes into a charge cloud with energy $E_X - E_{\text{Si K}\alpha}$. The Si K$\alpha$ line in the instrument spectrum is due to fluorescence photons that travelled far

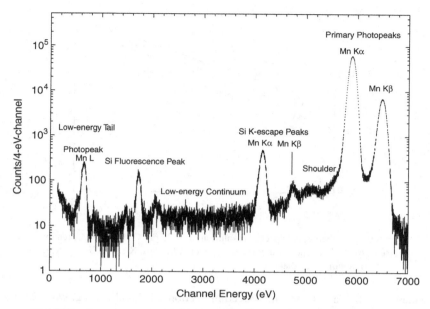

Fig. 3.10   An example spectrum demonstrating the principal components of the CCD spectral response. The source spectrum has lines of Mn L (0.7 keV), Mn Kα (5.9 keV) and Mn Kβ (6.4 keV). The Si K fluorescence and escape peaks and the low-energy features are due to the instrument. Source: Science Instrument Calibration Report for the AXAF CCD Imaging Spectrometer (ACIS)

enough from the original interaction site to be independently detected. The Si K-escape peak, at $E_X - E_{Si\,K\alpha}$, is due to the electron clouds that remain after the fluorescence photons "escape." The other features, such as the shoulder of the main photopeak and the low-energy continuum, are due to incomplete charge collection for events that interact in the gate and insulator structure or the channel stops.

3.3.3 Low-energy efficiency and back-illumination

Many astrophysically interesting problems require good low-energy detection efficiency, but most low-energy X-rays impinging on a CCD are lost to absorption in gate structures and the optical blocking filter. Two types of mitigating solutions have been attempted. In the first, the gates on the CCDs can be further thinned or can be "open" in which one gate is wider and has holes etched through to the oxide layer to reduce absorption. The XMM–Newton EPIC-MOS and Swift XRT detectors both use open-gate architecture. The other mitigating option is to use back-illuminated CCDs, which are currently part of the focal planes of the Chandra ACIS, XMM–Newton EPIC-pn, and the Suzaku XIS detectors.

A back-illuminated CCD starts out the same as a front-illuminated CCD. As the name implies, the CCD is flipped over such that the gate structures and channel stops are on the side facing away from the X-ray illumination. The back layer, which is now facing towards the X-ray illumination, is thinned so that the substrate is fully depleted. Instead of hundreds of nanometers of gate and insulator material between the active region and the incoming X-rays, there is only a thin surface dead layer.

By reducing the thickness of dead layers, the low-energy quantum efficiency (QE) of a back-illuminated CCD is much higher than that of a front-illuminated CCD. If thinning reduces the size of the active depleted region, such as for the ACIS back-illuminated CCDs, the high-energy QE is somewhat reduced. This is not the case for the XMM–Newton EPIC-pn CCDs, which, while back-illuminated, are quite thick and retain good high-energy QE. The resolution of back-illuminated CCDs also tends to be worse due to increased noise and charge-transfer inefficiency from defects in the surface. The difference in spectral resolution performance between front-illuminated and back-illuminated CCDs has reduced as the technology has become more mature. For example, the spectral resolution of the back-illuminated CCD on Suzaku is quite similar to the front-illuminated CCDs.

3.4 Detector features

There are a number of features of CCD detectors as used in X-ray astronomy that are important for the observer to understand in order to interpret data properly. Each observatory has specialized data analysis tools and calibration products which minimize the effect these features have on images, spectra, and lightcurves, but observers still need to be aware of the residual effects that may remain. The reader is encouraged to explore the web pages and analysis guides provided by each observatory for further mission-specific guidance.

3.4.1 Bright sources: pileup and readout streak

If two or more photons interact within a few pixels of each other before the image is read out, the event-finding algorithm may identify them as a single event. This can result in a higher inferred energy for the detected events, reduction of their total number, spectral hardening of continuum sources, and distortion of the point spread function of the telescope. There are models and algorithms which can be used to correct for pileup in analysis, but it's usually best to set up the observation with the goal of minimizing pileup. Pileup is a function of the source count rate per pixel so it is not just a function of the source flux and instrument effective area, but also of the relative sizes of the CCD pixel and the telescope point spread function.

A related effect is the readout streak caused by out-of-time events. As discussed in Section 3.2.5, X-ray CCDs are operated in photon counting mode. Accurate spectroscopy requires single photon events, in turn requiring short frame times of a few seconds or less. Optical CCDs used for astronomy generally integrate over much longer frame times, hundreds to thousands of seconds, to collect many photons in each pixel. In addition, they usually have a mechanical shutter which shields the CCD from light while the image is being transferred and read out. Astronomical X-ray CCD cameras generally do not have a shutter, because the much shorter frame time would require shutters capable of millions of cycles of reliable operation. This means that while the image is being transfered to the readout, events can still reach the CCD and be detected. The row number, or CCD y-coordinate, assigned to the out-of-time events, will be incorrect, as will any position-based calibration. Streak events can be modeled and removed from images for esthetic reasons, but can also be useful for analyzing bright sources since the time resolution is much higher with essentially no pileup.

3.4.2 Radiation damage and charge-transfer inefficiency

Charge-transfer inefficiency or CTI[1] is a common problem for CCDs. As charge is transfered across the CCD from the location of the X-ray photon to the readout amplifiers, charge can be lost to trapping sites in the detector. This leads to degraded and position-dependent gain, detection efficiency, and spectral resolution. These charge traps are caused by defects created in the manufacturing process or by radiation damage. CTI is a complicated function of the density of the charge traps, the trap capture and re-emission time constants (which are temperature-dependent), and the occupancy of charge traps (which depends on the particle background level and optical loading).

All CCDs have some level of CTI, particularly after many years in the radiation environment of space. Spacecraft operations and scheduling are designed to minimize the exposure of the instruments to high radiation environments where possible. During passages through the Earth's radiation belts, instruments are

---

[1] Some documents refer to charge transfer efficiency or CTE $= 1-$ CTI.

shut down and protected. In addition to scheduled protection measures, missions actively monitor the radiation environment and shut down during potentially damaging solar events.

While all energetic particles can cause radiation damage to CCDs, low-energy protons (hundreds of kiloelectron-volts) can be a particular issue because they scatter off grazing incidence mirrors in the same way as X-rays and thus are concentrated toward the focal plane. Low-energy protons are doubly problematic because they preferentially create charge-trapping sites in the buried transfer channel in front-illuminated CCDs. The Chandra ACIS CCDs were exposed to such protons during unprotected radiation-belt passages early in the mission. This produced a large increase in the CTI of the front-illuminated CCDs but no measurable increase in the CTI of the backside-illuminated CCDs, due to their much deeper-buried transfer channel. Missions in low Earth orbit do not encounter significant soft-proton fluxes but still suffer radiation damage from higher-energy particles, particularly during passages through the South Atlantic Anomaly (SAA).

There are a number of techniques in use to mitigate CTI. In general, lowering the focal-plane temperature can reduce CTI. This is a reason that instruments operate as cold as possible, at $-90\,°C$ and below. Since the performance degradation is a function of the number of transfers, some observations can be planned such that the target is imaged closer to the CCD readout. This is often done for Chandra observations that use the transmission gratings, where the spectrum is dispersed along CCD rows and can be moved from the aimpoint hundreds of pixels closer to the readout register. Some CCDs have the capability to inject a controlled amount of charge during the transfer, which temporarily fills the electron traps that cause CTI. This type of charge injection is routinely used on the Suzaku XIS detector and is quite successful at reducing CTI. In addition, some of the performance degradation due to CTI can be corrected and removed post-facto with specialized analysis software. Since charge loss is a stochastic process, post-facto mitigation techniques cannot recover all the lost performance.

3.4.3 Contamination

As discussed above, for best performance, CCDs must be operated cold. Each instrument uses a combination of active and passive thermal control to reach operating temperatures as low as $-120\,°C$. The CCDs are therefore often the coldest surfaces in the spacecraft and are in danger of accumulating contamination from outgassing materials. The contaminant acts as an additional absorbing layer. At low energies below 1 keV, contamination strongly affects the detection efficiency, while at higher energies it is relatively unimportant. Some degree of contamination is a common problem among the existing X-ray missions. For example, Chandra's ACIS detector has developed a layer of a hydrocarbon contaminant on the optical blocking filter. The thickness of this layer has increased throughout the lifetime of the mission.

3.4.4 Background

As well as X-ray events from the source of interest, an observation can also include those from other sources, both point and diffuse, resolved and unresolved, scattered solar X-rays, and events due to the charged-particle background (see also Section 8.2). Defining which events are part of the background depends on the goals of the observation. The spectrum and temporal variability of the charged-particle background depends on the radiation environment of the spacecraft and therefore on the spacecraft orbit. All current and past missions fit into two orbital categories, low Earth and high Earth.

Chandra and XMM–Newton are in highly elliptical orbits that dip into the Earth's radiation belts at perigee. Outside of the radiation belts, the particle background can be separated into two components; the slowly evolving quiescent background and short-lived flares (see Section 8.2.1). The quiescent background is primarily due to cosmic rays and is anti-correlated with the solar cycle, such that it is highest at solar minimum and lowest at solar maximum. There can also be flares in the particle background which can last from minutes to hours. These flares are usually lower-energy particles than those that cause the quiescent background and affect back-illuminated more than front-illuminated CCDs. There is some correlation of these flares with solar activity and geomagnetic conditions and they are believed to be due to geomagnetic protons with energies around 100 keV. In a few cases, when the radiation environment is particularly extreme, higher background rates can also be seen just before or just after the instruments are shut down for radiation-belt passages.

The remaining spacecraft are in low Earth orbit and are partially protected from cosmic rays by the Earth's geomagnetic field. This means the quiescent background level is much lower in low Earth orbit than for Chandra or XMM–Newton. Instruments are generally shut down while passing through the SAA due to the much higher radiation conditions. The background rate varies throughout the orbit as a function of the geomagnetic cut-off rigidity, a measure of how well the Earth's geomagnetic field shields the spacecraft from charged particles.

If an observation requires low and stable background, the event list can be filtered to remove times contaminated by high background. The quiescent background is reduced by event-morphology (grade) filtering, however this is less effective for back-illuminated CCDs where cosmic-ray events better mimic the appearance of X-rays. Additional background reduction is possible using specialized modes, such as the ACIS very faint mode. In this case, the pixel values in the outer 5×5 pixel ring around the event island are used to better identify large diffuse charged-particle blooms. Otherwise, the background can be modeled, or estimated and subtracted.

3.4.5 Micrometeoroid damage

An additional danger to CCDs as focal-plane instruments of X-ray observatories is damage from micrometeoroids. Micron-sized particles incident on grazing-incidence optics will preferentially scatter along the mirror in the direction of the focal plane. The consequences of an impact on a CCD vary. Initially a bright flash of light may fill memory buffers or saturate telemetry resources. The long-term damage can range from a few new hot pixels or columns (see Section 3.4.6) to a pin-hole in an optical blocking filter to the loss of an entire readout node or CCD. The likelihood of micrometeoroid impacts on a specific observatory depends on the spacecraft orbit as the particle flux in low Earth orbit is about twice that seen in higher orbit. The likelihood also depends on the mirror area and geometry, with higher probability for the largest collecting areas and lowest grazing angles. At the time of writing, suspected micrometeoroid impacts have been reported on the XMM–Newton EPIC, Swift XRT, and Suzaku XIS detectors. While Chandra and XMM–Newton are in similar high Earth orbits and should be experiencing similar particle fluxes, XMM–Newton has had multiple micrometeroid events while Chandra has had none, most likely due to XMM–Newton's larger effective area and shallower grazing angles.

3.4.6 Pixel defects

Radiation damage or manufacturing defects can cause pixels or even entire columns to have anomalously high dark current. Dark current is caused by random generation of electron–hole pairs in the depletion region. At the low temperatures and fast readout rates used for most X-ray CCDs, dark current is negligible except for these anomalous pixels. These warm or hot pixels can regularly exceed the event threshold and cause spurious events. Extreme cases may be removed on-board to avoid wasting valuable telemetry bandwidth, otherwise they should be filtered and removed in data analysis. Hot pixels are strongly correlated with CCD temperature, and so are more important for ASCA ($-60\,^{\circ}$C) and Suzaku ($-90\,^{\circ}$C) than Chandra and XMM–Newton ($-120\,^{\circ}$C).

In particular, ASCA was affected by flickering pixels, hot pixels that appear and disappear with a duty cycle of minutes to days. Because they occur with lower frequency, flickering pixels are more difficult to detect and remove.

Cosmic-ray afterglow events occur when a cosmic ray produces an especially large amount of charge that fills surface states with long time constants. These traps will release the charge slowly, over many minutes. The result looks like a temporary hot pixel or group of pixels that fades away over time. Since the phenomenon is not associated with a particular physical pixel, standard filtering used for pixel defects is not useful. Specialized software is needed to search event lists for the signature of these events.

3.5 Future X-ray imaging detectors

Photon-counting X-ray CCDs are the workhorses in almost all the currently operated X-ray observatories. Although microcalorimeters (see Section 2.7), in development for missions such as Astro-H and IXO, have much better spectral resolution than CCDs, at least for the near future they are unlikely to offer large focal-plane arrays or good low-energy QE. Large megapixel imaging detectors will be required to read out future grating spectrometers and also for wide-field imaging. Technologies are being explored to provide better low-energy response, radiation tolerance, and time resolution as well as faster readout rates to accommodate the large effective-area mirrors of future observatories.

The most interesting improvements may come from faster readout detectors. Most current imaging detectors have readout times of order seconds per frame. Faster readout offers many advantages, including the obvious one of reduced pileup, but other less obvious ones as well. Faster readout yields lower dark current per frame so that the detector can operate at higher temperatures. It also means less optical contamination per frame so thinner optical blocking filters can be used, giving better low-energy efficiency. Highly parallel readout with an amplifier on each column or active pixel sensors are under development and may yield orders of magnitude increases in readout rate and signal processing.

Now, with almost twenty years of operating experience in space applications, CCDs have become the workhorses of X-ray astronomy. Although new and exciting technologies are being developed that will improve upon aspects of CCD performance, CCDs still have many qualities that recommend their inclusion in future X-ray observatories.

4

Data reduction and calibration

KEITH A. ARNAUD AND RANDALL K. SMITH

The next five chapters cover the practical details of data analysis in X-ray astronomy. In this chapter we will consider what it takes to get the data to the stage of performing scientific analysis, which will itself be described in the following chapter. We will then survey the archives, catalogs, and software available, follow that with a short discussion of statistical issues peculiar to X-ray astronomy, and finish with the special considerations required when analyzing observations of sources which occupy the entire field-of-view (FOV).

We start with a description of the standard files, their contents, and the processing pipeline which produces them, as well as the initial data-reduction steps typically performed by the individual scientist.

4.1 The event file

All X-ray detectors measure individual photons. This is in contrast to many instruments for longer wavelengths, which measure integrated flux. The reason for this difference is that X-ray photons have relatively large energies, so single ones can be easily detected, but have relatively low fluxes, so they are easy to count. The basic data structure is thus a list of detected events, each of which has a set of attributes. Current X-ray instrumentation typically measures the position the X-ray arrived on the detector, the time of arrival, and some attribute which relates to the energy of the photon. Polarization is harder to determine and the first sensitive X-ray polarimeter is due to be launched on the GEMS satellite in 2014. In addition to these basic attributes, most event lists will also include other quantities which can be used to discriminate good events from background, such as grade (or pattern) in CCDs (see Section 4.1.2 and Section 4.1.5).

Table 4.1   *Attributes for each X-ray in a Chandra ACIS event file*

Column name	Min	Max	Description
TIME			Spacecraft time (s)
CCD_ID	0	9	CCD chip
NODE_ID	0	3	Section of chip
EXPNO			Readout number
CHIPX	1	1024	Chip position
CHIPY	1	1024	Chip position
TDETX	1	8192	Detector position
TDETY	1	8192	Detector position
DETX	0	8192	Detector position
DETY	0	8192	Detector position
X	0	8192	Sky position
Y	0	8192	Sky position
PHAS	−4096	4095	Vector of PHA for grade
PHA	0	36855	Total PHA for event
ENERGY	0	1000000	Estimated event energy
PI	1	1024	Gain-corrected PHA
FLTGRADE	0	255	Event Grade
GRADE	0	7	Event Grade
STATUS			Status flags

Table 4.1 lists the attributes provided in a Chandra ACIS event file. These include the time, the position in several coordinate systems, energy measurements, the event grade and a status value which can be set to indicate problems.

Practically all X-ray astronomy data is stored in FITS[1] format (Wells *et al.*, 1981). An event file typically consists of the empty primary image extension required by the FITS standard, followed by a binary table extension with the event list, then a number of subsidiary extensions containing information required for analysis of the event list.

Other information about the status of the detector and satellite are stored in housekeeping files, usually at a fixed cadence. Table 4.2 shows some of the housekeeping parameters which are important for selection of good data from the Suzaku XIS device.

The raw event attributes are usually not sufficient for scientific analysis so they are processed to create refined attributes. This is normally done in an automated pipeline run by the mission data center. The major steps are as follows.

---

[1] http://fits.gsfc.nasa.gov

Table 4.2 *Example housekeeping attributes for the Suzaku XIS*

Column name	Description
TIME	Spacecraft time (s)
AOCU_HK_CNT3_NML_P	Whether attitude control is in pointing mode
ANG_DIST	Angle between instantaneous pointing and the mean
Sn_DRATE	Telemetry rate for sensor n
SAA_HXD	Whether satellite is in SAA
T_SAA_HXD	Time since last SAA passage
ELV	Angle between pointing direction and Earth's limb
DYE_ELV	Angle between pointing direction and the sunlit limb of the Earth

4.1.1 Calculation of sky position

The position initially associated with an event is that on the detector, or perhaps section of the detector. As an example, consider four CCDs used together as a single detector. The first step is to convert the raw detector position (such as "Event on CCD 3, position 431, 617") to one in a coordinate system fixed in the focal plane ("Event at position 3183, 2184"). This conversion requires knowledge of the pixel sizes and orientations of the detector sections. Except in unusual circumstances, this conversion will not vary with time. The new positions are referred to as detector coordinates (conventionally DETX, DETY) and are usually defined looking down on the focal plane from the sky. The next step, converting to sky coordinates, requires a knowledge of the direction that the focal plane was pointing when the event was detected. This is supplied by an auxiliary star-tracker telescope, which will itself not be pointing in precisely the same direction as the focal plane. The conversion from detector to sky coordinates may be time-variable; satellite structures can flex due to changing thermal stresses, and in some cases satellites may deliberately dither about the nominal pointing location. The sky coordinates are calculated for a tangent plane approximation to the celestial sphere, with the "normal" direction of the plane defined as the nominal pointing direction for the observation (which is stored as the keywords RA_NOM and DEC_NOM). The positions (X, Y) recorded in the event file are in units of pixels and can be converted to right ascension and declination using the associated World Coordinate System (WCS) keywords.[2] Figure 4.1 shows the same XMM–Newton observation in detector and sky coordinates. In this case, the conversion from detector to sky coordinates requires a rotation and a flip.

---

[2] http://fits.gsfc.nasa.gov/fits_wcs.html

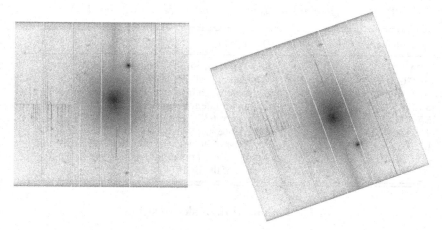

Fig. 4.1   The cluster Abell 1795 observed with the XMM–Newton EPIC-pn
detector (ObsID 0097820101) in detector (left) and sky (right) coordinates

An issue specific to Chandra is that, because of its superb telescope, the
detector pixel sizes are comparable to the spatial resolution. If all events were
placed at the center of detector pixels the resulting image would have a gridded
appearance. Initially, Chandra processing avoided this by randomly assigning
each event a position within its pixel. In December 2010, with the release
of version 4.3 of the Chandra software (CIAO 4.3), this was changed to use
subpixel repositioning, which reduces the effective pixel size by using event
shapes (Li *et al.*, 2004).

4.1.2 Calculation of grade

For CCDs, an event is defined as a pixel whose charge exceeds some predetermined threshold. For each event the charges in the surrounding 3×3 or 5×5
pixel regions are also recorded (see Section 3.2.5). The grade (or pattern for
XMM–Newton) is a number characterising the pattern of pixels whose charges
are above some value. For instance, if only the central pixel has charge above
the cut-off then the grade is 0. The range of values and their meanings vary
from instrument to instrument. Figure 4.2 illustrates grades for the Chandra
ACIS.

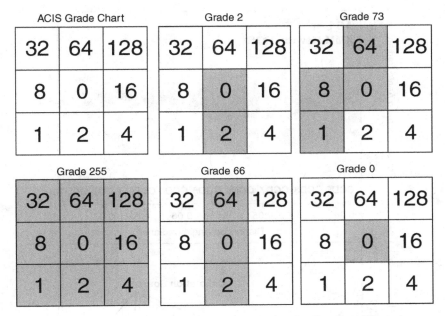

Fig. 4.2 Illustration of different CCD grades, using the Chandra ACIS grade system. A single event is defined as all contiguous pixels in a frame with charge above some minimum value. The event center (pixel "0") is taken to be the pixel with largest charge. Gray boxes indicate pixels with charge above the minimum. The grade is calculated by summing the numbers in each gray box. Some grades, such as 255, are automatically rejected as being almost certainly due to particle impact rather than an X-ray photon. Grade 0 events, also known as single-pixel events, have the best energy resolution, although double- and triple-pixel events are usually not much worse

4.1.3 Calculation of energy or wavelength

Non-dispersive spectrometers such as CCDs or proportional counters accumulate charge for each X-ray detected. For historical reasons this amount of charge is called the PHA (for pulse-height amplitude). The conversion between PHA and energy may not be strictly linear and can vary with time or position on the detector. Any time or position variability is corrected and the result stored as PI (for PHA invariant). The relationship between PI and the energy of the photon is encoded in the response matrix, described in Section 4.5.2.

For CCDs, the charges in the 3×3 or 5×5 pixel regions are stored in a vector, PHAS. The total charge for the event is the sum of PHAS charges for those pixels above the threshold value used when calculating the grade.

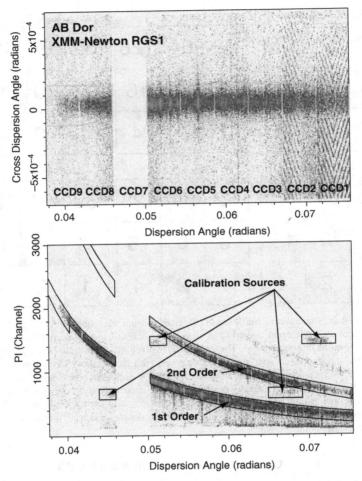

Fig. 4.3 [Top] XMM–Newton RGS observation showing the image of the dispersed spectrum of the star AB Dor. Although some emission lines (visible as thin vertical lines) can be seen, all of the orders are overlaid upon each other. The cross-hatching seen in CCDs 1 and 2 is an artifact of the CCD readout. [Bottom] Same, but plotted with the measured CCD energy for the *y*-axis. This creates the "banana plot" that shows the 1st and 2nd order clearly, with a faint third-order curve visible as well. The boxed regions show X-rays from small radioactive calibration sources built into the gratings themselves to monitor any gain and efficiency variations

Table 4.3   *Definition of spacecraft MJD offset for*
active missions

Mission	MJD offset
Chandra	50814.000000000000
MAXI	51544.0007428703703703700
RXTE	49353.000696574074
Suzaku	51544.0007428703703703700
Swift	51910.000742870370
XMM–Newton	50814.000000000000

Dispersive spectrometers such as reflection or transmission gratings deter-mine wavelength from the distance in the dispersion direction between the un-dispersed position of the source (the so-called "zero order") and the posi-tion of the photon (see Section 1.4). However, this distance may not be a unique measure of the wavelength because of overlap between diffraction orders. This degeneracy can be broken when the detector itself has energy-resolution (e.g. a CCD) since each event can be placed on a two-dimensional plot with axes of the dispersion distance and the PI (this is somewhat analogous to the cross-dispersion used in optical echelle spectroscopy). Each order of the grating lies along a different curve in this plot, allowing both order separation and efficient background exclusion (see Figure 4.3).

4.1.4 Calculation of time

The event time is recorded on-board, relative to the spacecraft clock. This is then converted to a time at Earth and recorded in seconds relative to the modified Julian date (MJD)[3] listed for each operating mission in Table 4.3. For most purposes this is adequate, but for observations of sources with high time variability or precise ephemerides the time should be converted to Solar System barycenter to eliminate effects due to the Earth's orbit. This is not done in the automatic processing because the precise information required about the satellite orbit is typically not available until several weeks after the observation and most missions aim to process data faster than that.

Those missions carrying detectors such as CCDs (see Chapter 3), which accumulate events within a frame time and then read them out, need a rule for assigning a time to the events within each frame. Chandra and Suzaku use

---

[3] Julian date is the number of days since noon on January 1, 4713 BCE. MJD is Julian date − 2400000.5

the middle of the frame, Swift the start of the frame while XMM–Newton randomly assigns a time within the frame to each event.

4.1.5 Non-X-ray background rejection

Most of the non-X-ray background in X-ray detectors is due to the interaction of energetic particles (cosmic rays) with the detector. The signal due to these particles often differs in some way from that due to X-rays. For instance, in CCDs it will produce a different pattern of charge in neighbouring pixels; see Figure 4.2 for an example using the Chandra ACIS system. Different instruments have different definitions of grades and choices about which can be assumed to be background. The signal-to-noise ratio can be improved by rejecting events which are likely to be background. Processing systems are usually set to reject types of events that are always background but leave the individual researcher the option to select further if they are willing to remove more background events at the cost of rejecting some true X-rays. However, not all possible choices will have been calibrated so an unusual choice may require that the observer do some calibration work themselves.

4.2 Looking at the data

The first thing to do when acquiring a new data set is always to look at it in as many ways as possible. Do not assume that the automated processing has worked correctly. For some missions, a member of the processing team will have looked at the output, but for other missions, with smaller budgets, any checks will be automated and cursory. Processing may need to be rerun to use the latest calibration information. Individual missions provide instructions on how to do this.

4.2.1 Background flares

All missions have time-variable backgrounds but some are worse than others. Spacecraft in deep orbits, such as Chandra and XMM–Newton, spend most of their time outside the protection of the Earth's magnetosphere and are vulnerable to particle storms from solar flares. These can be seen by making a time series of event rates as shown in Figure 4.4. For point sources observed on axis with Chandra, background flares are not a concern because the extraction region is so small it contains effectively no background events. For all other cases, however, unless the source is very bright, excluding times of background

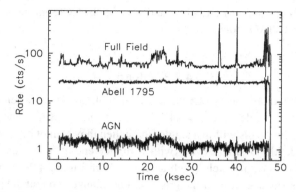

Fig. 4.4 Count rates from an XMM–Newton observation containing a cluster (A1795) and an AGN. Short flares can be seen as well as a longer period of enhanced background in the middle of the observation. This observation is discussed further in Section 5.2.5

flares from the data being analyzed will improve the signal-to-noise ratio. For sources with low surface brightness, doing this exclusion correctly is vital (e.g. Markevitch *et al.*, 2003). Some missions have specific software designed to assist with this process; see Chapter 8 for more details.

4.2.2 Hot spots, bad rows, flickering pixels, and afterglows

Detectors often have electronic hot spots or bad rows which produce apparent signals which are neither X-ray nor background events (see also Section 3.4.6). In some cases, events from these regions may be excluded on-board the spacecraft in order to reduce the telemetry load. If not, they are generally excluded during processing. Either way, a list of the excised data is stored either in an extension of the event file (e.g. the XMM–Newton BADPIXnn extensions) or a separate file (e.g. the Chandra bpix1 file).

A more difficult problem for CCDs involves so-called flickering pixels which are sporadic. The prevalence of these depends on the temperature at which the instrument is running. Chandra runs cold enough that flickering pixels are not a problem. Other missions have pixel sizes much smaller than the telescope point response so anomalous pixels can be identified by comparing them with their surrounding brethren: a single pixel which is much brighter than adjacent pixels cannot be due to a real X-ray source. This statistical cleaning is performed by the HEAsoft program `cleansis` and the XMM–Newton SAS program `emevents`.

An additional effect observed in the Chandra ACIS instrument is for a pixel to show a signal for a few contiguous frames (i.e. a few tens of seconds). This "afterglow" is due to cosmic-ray interactions with the CCD and can be removed using standard CIAO tools.

4.2.3 Pileup

Integrating detectors such as CCDs are vulnerable to pileup when observing bright sources (see also Section 3.4.1). If two photons hit the detector in the same or adjacent pixels within a single frame then they cannot be distinguished and will look like a single event with a higher energy and, perhaps, more complicated pattern of pixels (such as grade 73 in Figure 4.2). This has two possible consequences. Either the grade will be changed so that the processing rejects both events or the two events will appear as a single event with the combined energy. An image of a bright source can appear to have a hole in its center as piled-up events are incorrectly rejected as background (Figure 4.5 [top]). The spectrum will be distorted as events are removed from lower energies and a smaller number added at higher energies. Figure 4.5 [bottom] shows that increasing the input count rate reduces the events registered at energies below 2 keV while increasing those above.

Pileup significantly impacts imaging, spectral, and timing analysis, and the best strategy is to plan the observation to reduce its importance by choosing instrument and operating modes appropriately. For example, if only a small section of the CCD is used, then the frame time can often be decreased. If this is not possible (e.g. because the full FOV is required) or adequate (e.g. pileup still occurs even with faster readout), then the spectroscopic effects can be minimized by using events only from the wings of the point response or from the out-of-time events. If none of these options are available and pileup is moderate, it can be explicitly modelled in spectral analysis (Davis, 2001b).

4.3 Selecting events of interest

4.3.1 By region

Any instrument with spatial resolution will likely require filtering to select the region of interest. Region filters consist of a combination of individual shapes (see Table 4.5), each of which can specify either that its constituent pixels should be included or excluded. Region filters can be defined either in a text file, FITS file (Rots and McDowell, 2008), or sometimes directly as an argument to software. Unfortunately, there are several different text formats in use. These are similar, but not identical, so it is important to make sure

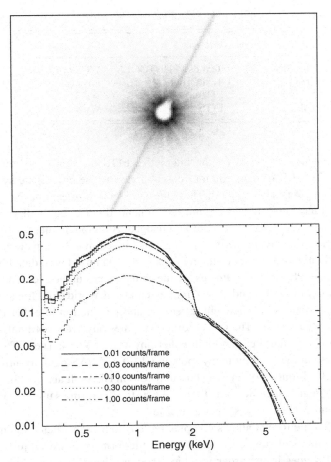

Fig. 4.5 [Top] Chandra ACIS observation of the bright source Hercules X-1. The hole in the center of the source is due to piled-up events which are rejected because of bad grades. Also visible are the streaks due to events detected while the detector is being read out and the shadows of the mirror support structure. [Bottom] Chandra ACIS simulated spectra showing the effect of increasing pileup. As the input count rate, and hence pileup, is increased the observed spectrum decreases at low energy while increasing at higher energies

the correct format is used for the software doing the subsequent filtering. The two main text formats are DS9/FUNTOOLS and CIAO, both of which can be produced by the image and data visualization program SAOImage DS9.[4]

[4] http://hea-www.harvard.edu/RD/ds9

Table 4.4   *Region file formats used*

Software system	File type	File units
HEAsoft	Text ds9/FUNTOOLS	WCS or Physical
CIAO	Text CIAO	Physical
	FITS	WCS
SAS	FITS	WCS

The region files produced by the interactive FITS file editor fv[5] are in the DS9/FUNTOOLS format. Another complication is the coordinate system in which the regions are specified. There are three possibilities: image, physical, or WCS. Image coordinates are not recommended since they will vary if the pixel binning (e.g. 1×1, 2×2, etc.) is changed. A region in image coordinates created when displaying the data with pixels binned in a 2×2 pattern will be offset and only one quarter of the required size when used with data binned at a 1×1 scale. Physical coordinates are the actual values in the event file (e.g. X and Y for sky or DETX and DETY for detector) and World Coordinate System (WCS) coordinates are the right ascension and declination in either decimal or sexagesimal format. The CIAO format assumes physical coordinates while ds9/Funtools format files can be in either physical or WCS. The FITS format for regions can be used by many tools in CIAO and SAS but currently (as of version 6.10) is only written as an output record in HEAsoft and cannot be used as an input region specification. Table 4.4 lists the file formats used by different systems and Table 4.5 the shapes available.

The FITS file format allows complex regions to be constructed by intersections and unions of simple shapes. The text file formats, however, just process individual shapes in the order in which they are listed. If an exclude region is followed by an include region which overlaps it then the intersection will be included. However, if the regions are listed in the opposite order then the intersection will be excluded. Figure 4.6 shows an example of two include circular regions and an exclude circular region. The left-most include region is given first in both cases but the order of the other include region and the exclude region are swapped. If the first region in the list is an exclude then a starting include of the entire field is assumed.

Although filtering by spatial regions enables fast and powerful analysis, the number of different formats available, and the limited power of some interpreters, means that the first step after applying a region filter should be

---

[5] http://heasarc.gsfc.nasa.gov/lheasoft/ftools/fv

Table 4.5 *Region shapes*

Shape	Arguments	Description	HEAsoft	CIAO	SAS
Point	(X,Y)	One-pixel square region	✓	✓	✓
Line	$(X1,Y1, X2,Y2)$	One-pixel-wide rectangle from $(X1,Y1)$ to $(X2,Y2)$	✓	X	✓
Polygon	$(X1,Y1, X2,Y2, \ldots)$	Polygon with vertices $(X1,Y1)$, $(X2,Y2)$,...	✓	✓	✓
Rectangle	$(X1,Y1, X2,Y2, A)$	Box with corners $(X1,Y1)$, $(X2,Y2)$ rotated by A	✓	✓	✓
Box	(X,Y,W,H,A)	Box with center (X,Y) width W, height H and rotation A	✓	X	✓
Diamond	(X,Y,W,H,A)	Diamond with center (X,Y) width W, height H and rotation A	✓	✓	✓
Circle	(X,Y,R)	Circle with center (X,Y) and radius R	✓	✓	✓
Annulus	(X,Y,Ri,Ro)	Annulus between Ri and Ro centered on (X,Y)	✓	✓	✓
Ellipse	(X,Y,Rx,Ry,A)	Ellipse center (X,Y) semi-major axes Rx, Ry and rotation A	✓	✓	✓
Ellipannulus	$(X,Y,Rix,Riy,Rox,Roy,Ai,Ao)$	Like Annulus region but with ellipses	✓	✓	X
Boxannulus	$(X,Y,W1,H1,W2,H2,A)$	Like Annulus region but with boxes	✓	✓	X
Sector	$(X,Y,A1,A2)$	Region center (Xc,Yc) between angles $A1$ and $A2$	X	✓	X
Pie	$(X,Y,Ri,Ro,A1,A2)$	Region center (Xc,Yc) between radii Ri, Ro and angles $A1$, $A2$	✓	✓	✓
Panda	$(X,Y,A1,A2,1,Ri,Ro,1)$	Same as Pie region	✓	X	X
Epanda	$(X,Y,A1,A2,1,Rix,Riy,Rox,Roy,1,A)$	Like Panda region but with ellipses	✓	X	X
Bpanda	$(X,Y,A1,A2,1,Wi,Hi,Wo,Ho,1,A)$	Like Panda region but with boxes	✓	X	X

All angles are measured in degrees anti-clockwise from the x-axis.
HEAsoft uses Pie as a synonym for Sector while SAS uses Sector as a synonym for Pie.
SAS defines W and H as the half-width and half-height in contrast to HEAsoft and CIAO where they are the full width and height.

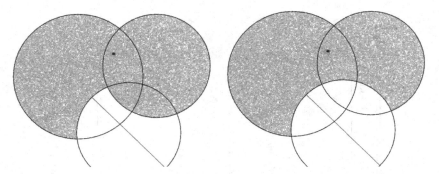

Fig. 4.6 Selected area when the bottom exclude circle is listed second (region 1 – left) or third (region 2 – right)

Region 1	Region 2
circle(3968.5, 3921.5, 288.2)	circle(3968.5, 3921.5, 288.2)
−circle(4192.5, 3661.5, 231.2)	circle(4316.5, 3969.5, 238.8)
circle(4316.5, 3969.5, 238.8)	−circle(4192.5, 3661.5, 231.2)

to check that the correct region was selected. Even simple region files with coordinates in WCS units can lead to difficulties. For example, the region file used to extract an on-axis point source with Chandra will be far too small to use with Suzaku data, since Suzaku's point-source response is much larger than Chandra's.

4.3.2 By time interval

The time intervals during which events are accumulated are referred to as the good time intervals (GTIs). The GTIs for an observation are listed in their own extensions in the events file. For instruments with multiple detectors that are not read out simultaneously (such as Chandra ACIS), there is a GTI extension for each detector. The extension(s) typically have names such as STDGTI, GTI, or GTI# where # is a number. The GTIs comprise lists of start and stop times, given in units of spacecraft time (see Section 4.1.4). The reference time is given in modified Julian date by the keyword MJDREF or the pair of keywords MJDREFI (for the integer portion of the offset) and MJDREFF (for the fractional portion).

All the major software packages provide ways of specifying additional GTI filters either by command line or graphical user interface (GUI). Some tools will convert from UT or MJD to spacecraft time, others require the user to do this themselves.

4.3.3 By phase

For sources with known periods (such as X-ray pulsars) it may be useful to extract events from a particular phase. Some software allows this to be done directly, otherwise it is simple to create a new column in the FITS file called PHASE, calculate this, then filter on it.

As an example, the high-mass X-ray binary X Persei exhibits an 837-second periodicity in X-rays (Delgado-Martí *et al.*, 2001), slow enough to be easily detected even with CCDs. Adding the new column that lists the phase can be done in two steps, shown here using the CIAO tool dmtcalc:

```
unix% dmtcalc evt.fits tmp.fits "GPH=(time-TSTART)/837"
unix% dmtcalc tmp.fits ph_evt.fits "PHASE=GPH-(long)GPH"
```

This results in a new file ph_evt.fits, which contains two new columns, GPH and PHASE. GPH increases by 1 for each phase that passes, while PHASE retains only the phase information. If an absolute value for phase 0 has been determined from other observations, this can be included as a fixed offset in the above equations. A similar operation can be done with the HEAsoft tool ftcalc.

When using complex filtering like this, care must be taken in later analysis that the assumptions underlying the standard calibration products are still applicable. Chandra dithers in a Lissajous pattern with a 1000-second period. A short segment of data from a source like X Persei, with its similar period, could beat against this dither pattern in unfortunate ways. For example, the two periods might coincide such that all the data around phase 0.2 occurred when X Persei happened to be positioned by the dither between two CCDs. As a result, the total flux might appear to drop precipitously at this phase because neither CCD detected any counts from it. The standard calibration includes a general scaling factor for sky positions that can fall off the CCDs due to dither, but does not assume that the user has extracted one particular phase that is more or less impacted.

4.3.4 By intensity/rate

It is often useful to select events based on the source (or background) event rate at the time of the event, such as when the source is bright or when the background rate is particularly low. The simplest method is to construct a GTI file that includes the desired times, then use time filtering. All of the standard X-ray packages include tools that enable the creation of arbitrary GTIs and their use for filtering. The advantage of this approach is that subsequent calibration tools can automatically calculate the total good time for the observation using information stored in the filtered FITS event file. As an example, here is intensity filtering using the HEAsoft program `xselect`:

```
xsel:SUZAKU-XIS0-STANDARD > extract curve

xsel:SUZAKU-XIS0-STANDARD > filter intensity 1.0-2.0

xsel:SUZAKU-XIS0-STANDARD > extract events
```

A Suzaku XIS event file has already been read in. The first command creates a lightcurve, the second command uses this lightcurve to generate a list of GTIs for when the count rate is between 1.0 and 2.0, and the third command makes a new event file using these GTIs.

4.3.5 By spectral channels

Filtering on spectral channel is often used when creating images or lightcurves, either to select a scientifically interesting energy band or to improve the signal-to-noise ratio. Spectral filtering is less often used when the final goal is a spectrum, since all spectral analysis tools can easily include and exclude any desired energy ranges. Filtering out the lowest and highest spectral channels can improve the signal-to-noise ratio as most instruments are less efficient at detecting X-rays at the low and high ends of their bandpass, and X-ray mirrors often lose area rapidly at the high-energy end as well. The background can also increase at the low end due to detector electronic noise and at the high end due to energetic particles (Figure 4.7). In most cases, there is no fixed recommendation for which energies should be eliminated, as a source that is extremely bright in soft or hard X-rays might still return useful information in a spectral region often dominated by background. We give typical bandpasses for many instruments in Appendix 4.

4.3.6 By grade

Section 4.1.5 described using event grade to reject the non-X-ray background. Grades which are always background are removed during standard processing; however, additional filtering can still be useful. The most common choice is to select only events with all the charge in a single pixel. This will reduce background at the cost of cutting the efficiency at higher energies, where X-rays

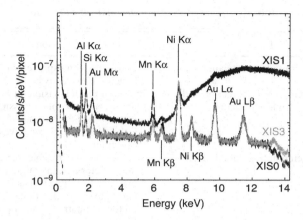

Fig. 4.7 Non X-ray background for the Suzaku XIS instrument. The identified peaks are due to fluorescent emission from materials in or near the detector. The upper line is for the backside-illuminated CCD XIS1, the others are for the frontside-illuminated detectors XIS0 and XIS3 (XIS2 is no longer operating due to an apparent micrometeoroid impact)

are more likely to spread their charge across multiple pixels. It is important to read the documentation for the instrument when doing grade filtering because not all choices may be calibrated.

4.3.7 By auxiliary criteria

Filtering on other criteria can also be useful. Most missions generate a time-series file of instrument and satellite housekeeping information. The most commonly used filters are listed in Table 4.6. For instance, Smith *et al.* (2007) improved their signal-to-noise ratio for an observation of extended emission with low surface brightness by excluding times when the pointing direction was within 10° of the Earth's limb and when the cut-off rigidity (a measure of magnetic field strength) was less than 8 GV.

4.4 Extracting analysis products

Most data analysis tools do not work directly with the event file but on subsidiary products. These can be thought of as taking the multi-dimensional event space (energy, time, position), projecting onto one or two dimensions, then binning. Such binning makes histograms of counts in one or two dimensions.

Table 4.6   *Common filters on auxiliary criteria*

Filter	Reason
Angle between pointing direction and Earth's limb	The extended atmosphere can absorb lower-energy X-rays.
Angle between pointing direction and Earth's day–night terminator	The sunlit Earth is visible in X-rays due to scattering of solar emission and fluorescence of the atmosphere. Also, some detectors are sensitive to optical contamination.
Angle between pointing direction and Sun	Solar X-rays may be able to scatter into the detector at small angles. In practice, the angle is usually determined by operational constraints on the orientation of solar panels to the Sun.
Cut-off rigidity (or McIlwain L)	The Earth's magnetosphere shields the spacecraft from high-energy particles. The cut-off rigidity and McIlwain L are measures of this protection.
SAA flag	The SAA is a region of the Earth's atmosphere above the South Atlantic that has intense particle bombardment. Some instruments are turned off in the SAA; others continue operating but the data are often not useful.
Time since last SAA passage	The SAA particle bombardment can generate short-lived radioactives which themselves provide background as they decay.

4.4.1 Image

Images are usually in sky coordinates (typically the X and Y columns in the event file) although detector coordinate (DETX and DETY) images can be useful to check for detector artifacts. For instance, Chandra has a deliberate wobble (dither) which means that any detector feature will be smeared out in the sky-coordinate image but will show up clearly in a detector coordinate image (see Figure 4.8 for an illustration of the effect of the Chandra dithering). Some instruments (e.g. the XMM–Newton EPIC) have a pixel size much smaller than the telescope spatial resolution. In this case, the image should be binned up. A rule of thumb is to choose a bin size about one third of the full width at half maximum of the spatial resolution.

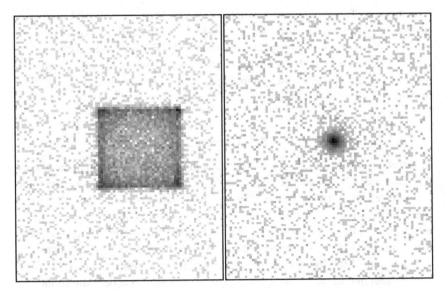

Fig. 4.8 A Chandra ACIS point source in detector (left) and sky (right) coordinates. The square appearance of the source in detector coordinates is caused by Chandra's dither pattern, designed to distribute flux over a 32 × 32 pixel region

Sometimes non-standard images are made by binning other columns in the event files. For example, to visualize source spectral variability use TIME and PI (energy) columns so each pixel contains a number of counts at a given time and energy. To be useful, such images require high signal-to-noise data.

4.4.2 Spectrum

The columns in the event file with energy information are PHA and PI. The former contains the raw information and the latter is corrected for time-dependent gain changes. CCD instruments usually also have a vector column PHAS that records the charge in each pixel of the 3 × 3 or 5 × 5 region that contains the event. The standard processing uses the PHAS to calculate PHA and PI as well as the grade (Section 4.1.2). Some files also contain an ENERGY column which is derived from PI and the detector gain. It should not be used for scientific analysis but can be plotted to get a quick estimate of energies of features such as lines or edges. Usually the spectrum should be made from the PI column, after ensuring the event file has been processed to use the most up-to-date calibration.

In the case of non-imaging detectors (e.g. RXTE PCA, Suzaku HXD) or high-resolution detectors (Chandra HETG, XMM–Newton RGS), the pipeline software itself will often create spectra automatically. For imaging spectrometers (such as CCDs) it is likely that the spectrum wanted will be from a region, not the entire image. Often a background spectrum is then created using another region.

4.4.3 Lightcurve

Lightcurves or time series are constructed using the TIME column in the events file. Care must be exercised with data from instruments such as CCDs with periodic readouts. Using a lightcurve binning which is not a multiple of the frame time will cause beating as the number of CCD frames included in a given bin changes. This may not be significant if the bin size is large compared to the frame time (which is usually ≤ 10 seconds), but can cause problems otherwise. For instance, if the lightcurve bin size is 1.5 frame times then alternate bins will contain one frame or two frames generating a spurious periodic signal.

4.5 Calibration

Standard processing in X-ray astronomy does not usually make products in physical units, free of all detector characteristics. Instead, information about the detector is included during the analysis step using calibration files. This section covers the main types of calibration file. Section 6.4 includes a discussion of how to find the right calibration file for a given observation.

Consider an exposure which produces the observed counts $C(X, Y, PI)$ in a given pixel and PI bin. These counts have come from the source flux $S(X_S, Y_S, E, t)$ photons/cm^2/s/keV/arcmin^2 for a given position on the sky (X_S, Y_S) at energy E and time t. S and C are related by:

$$C(X, Y, PI) = \int \int \int \int R(X, Y, PI, X_S, Y_S, E, t)$$
$$\cdot S(X_S, Y_S, E, t) \cdot dX_S \cdot dY_S \cdot dE \cdot dt \quad (4.1)$$

where R is called the instrumental response and in this definition has units of cm^2. R thus provides a measure of the chance of a photon from sky position (X_S, Y_S) with energy E at time t ending up as a count in pixel (X, Y) and PI bin PI. The following sections consider contractions of R for different products. For a more complete and formal discussion, see Davis (2001a).

4.5.1 For imaging

An image is created by summing over the PI bins so:

$$C(X, Y) = \sum_{PI} C(X, Y, PI)$$

$$= \int \int \int \int \left(\sum_{PI} R(X, Y, PI, X_S, Y_S, E, t) \right)$$
$$\cdot S(X_S, Y_S, E, t) \cdot dX_S \cdot dY_S \cdot dE \cdot dt$$

$$= \int \int \int \int R_{\text{image}}(X, Y, X_S, Y_S, E, t)$$
$$\cdot S(X_S, Y_S, E, t) \cdot dX_S \cdot dY_S \cdot dE \cdot dt \quad (4.2)$$

where the instrumental response for images is:

$$R_{\text{image}}(X, Y, X_S, Y_S, E, t) = \sum_{PI} R(X, Y, PI, X_S, Y_S, E, t) \quad (4.3)$$

This can usually be split into the point spread function (PSF) and the effective area, EA (often called Exposure Map):

$$R_{\text{image}}(X, Y, X_S, Y_S, E, t) = \text{PSF}(r, \theta, X_S, Y_S, E) \cdot \text{EA}(X_S, Y_S, E, t) \quad (4.4)$$

where $r^2 = (X - X_S)^2 + (Y - Y_S)^2$ and $\theta = \arctan((Y - Y_S)/(X - X_S))$. The PSF encapsulates information about the spatial resolution of the telescope as a probability distribution of the event position on the detector from a point source. X-ray telescopes are often described as having a particular fixed resolution in terms of their half-power diameter (HPD) – e.g. Chandra has a $0.5''$ resolution while XMM–Newton has a $15''$ resolution telescope – but this is only useful in an informal sense. In reality, an X-ray telescope's PSF is usually strongly energy-dependent and often position-dependent as well, degrading at higher energies and further from the optical axis. The PSF will also not be azimuthally symmetric, often appearing elliptical when the source is off-axis and usually with radial lines caused by X-ray scattering from support structures in the mirror itself (Figure 4.9). The exact appearance of these is highly dependent upon the details of the mirror manufacture.

The effective area or exposure map is the telescope area at (X_S, Y_S) given an X-ray of energy E at time t. This term is time-dependent because near the edge of the detector a sky pixel may only spend part of the time on the detector due to satellite pointing dither. This effect can be calculated using the telescope's "aspect history," which lists its pointing direction as a function of time. This

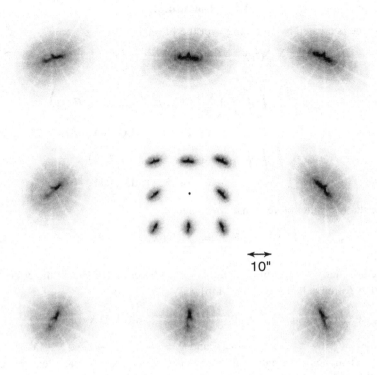

Fig. 4.9   The Chandra PSF as a function of position; the outer eight images are 10 arcmin off-axis, the inner eight are 5 arcmin off-axis, and the central dot is the on-axis PSF, 0.5 arcsec in diameter. Source: Chandra Proposers' Observatory Guide

information is normally provided as one of the auxiliary files. An example exposure map is shown in Figure 4.10.

To make a narrow energy band image in photons/cm^2/s/arcmin^2, such as might be used in combination with images in other wavelength bands, it is usual to assume that the PSF is a δ-function and use:

$$
\begin{aligned}
C(X, Y) &= \int \int \int \int \mathrm{EA}(X_S, Y_S, E, t) \cdot \delta(X_S - X) \cdot \delta(Y_S - Y) \\
&\quad \cdot S(X_S, Y_S, E, t) \cdot dX_S \cdot dY_S \cdot dE \cdot dt \\
&= \int \int \mathrm{EA}(X, Y, E, t) \cdot S(X, Y, E, t) \cdot \Delta A \cdot dE \cdot dt \qquad (4.5)
\end{aligned}
$$

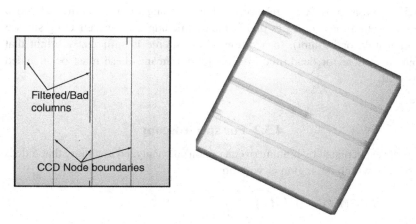

Fig. 4.10 [Left] Chandra ACIS instrumental effective area (exposure) map (at 1.5 keV) in detector coordinates, showing the node boundaries on the CCD and undesirable or bad columns that have been filtered. [Right] Same, after converting to sky coordinates and including the ACIS dither that smears out the bad columns

where ΔA is the area of an image pixel in arcmin^2. Assuming that S is not variable and the image is made for energy E_0 then

$$S(X, Y, E_0) = \left(\frac{C(X, Y)}{\Delta A}\right) \bigg/ \int E A(X, Y, E_0, t) \cdot dt \qquad (4.6)$$

Although the details of the PSF are not included, this method does include the effects of vignetting and removal of any bad pixels. If the image is over a wider energy band then it is necessary to suppose that S can be split into position-dependent and energy-dependent factors $S = S_{im} S_{spec}$ where S_{spec} is determined by spectroscopic analysis. If this is not an acceptable assumption then the bandpass must be narrowed. Then:

$$C(X, Y) = \int \int EA(X, Y, E, t)$$
$$\cdot S_{im}(X, Y, t) \cdot S_{spec}(E, t) \cdot \Delta A \cdot dE \cdot dt \qquad (4.7)$$

and assuming as before that S_{im} is not variable:

$$S_{im}(X, Y) = \left(\frac{C(X, Y)}{\Delta A}\right) \bigg/ \int \int EA(X, Y, E, t) \cdot S_{spec}(E, t) \cdot dE \cdot dt$$
$$(4.8)$$

A reasonable rule of thumb is that if the central energy of the band divided by the width of the bandpass is less than $\sim$4, then the latter method should be used. In addition, particular care should be taken for bandpasses around

1.5–2 keV, as most X-ray telescopes show a strong variation in their effective areas around these energies due to the atomic edges in the reflecting surface (often gold or iridium). In addition, if the source is sufficiently bright that pileup or detector dead-time effects are significant, these must be included as well.

4.5.2 For spectroscopy

Spectra are created by binning over a region and analyzed assuming that S does not vary within the region or with time.

$$C(PI) = \int \left(\int \int \int \int \int R(X, Y, PI, X_S, Y_S, E, t) \right.$$
$$\left. \cdot dX_S \cdot dY_S \cdot dX \cdot dY \cdot dt \right) \cdot S_{spec}(E) \cdot dE \qquad (4.9)$$

The response is usually split between a vector (the ancillary response file, ARF) with units of cm^2 and a matrix (the RMF), which is unitless. In this case, the equation is reduced to

$$C(PI) = T \int RMF(PI, E) \cdot ARF(E) \cdot S_{spec}(E) \cdot dE \qquad (4.10)$$

where T is the total good observing time,

$$RMF(PI, E) = \int \int \int R_{RMF}(X, Y, PI, E, t) \cdot dX \cdot dY \cdot dt \qquad (4.11)$$

and

$$ARF(E) = \int \int \int \int \int R_{ARF}(X, Y, X_S, Y_S, E, t)$$
$$\cdot dX_S \cdot dY_S \cdot dX \cdot dY \cdot dt \qquad (4.12)$$

If the source is diffuse and varies in surface brightness across the region then it may be appropriate to modify Equation 4.11 to give a source-flux-weighted RMF.

$$RMF(PI, E) = \int \int \int S(X, Y) \cdot R_{RMF}(X, Y, PI, E, t) \cdot dX \cdot dY \cdot dt$$
$$\Big/ \int \int S(X, Y) \cdot dX \cdot dY \qquad (4.13)$$

The RMF gives the probability of a photon of given energy ending up as a count in a particular bin in the spectrum. For a theoretical perfect detector this would be a diagonal matrix, as each energy would be mapped to a single channel in the detector. However, real detectors will always have some spread in their response, with some X-rays of energy E ending in channel I while others appear in channel $I - 1$ and others in channel $I + 1$, or even channel $I + 500$.

The total spectral response for energy E is:

$$\text{ARF}(E) \cdot \int \text{RMF}(PI, E) \cdot dPI \tag{4.14}$$

Some instrument teams define the RMF so that:

$$\int \text{RMF}(PI, E) \cdot dPI = 1.0 \tag{4.15}$$

for all energies but others include detector efficiency information in the RMF instead of in the ARF. As a practical matter, most missions separate the creation of these two calibration products with the telescope team responsible for the ARF while the detector team generates the RMF. In this case, the RMF usually includes detector quantum efficiency and the effect of any filters. For some instruments, the ARF and RMF are combined in a single matrix, usually referred to as the RSP (short for ReSPonse). Figure 4.11 shows one row on an example RMF (top panel) and ARFs for several current missions (bottom panel).

X-ray telescopes usually have PSFs with wide wings so a source region chosen to optimize signal-to-noise ratio may not include all the flux. The ARF may be defined relative to the flux enclosed within some radius. If the source region is different from this default then an aperture correction is necessary and this will usually be energy-dependent. Note that calibration uncertainties associated with ARF or RMF files are typically not taken into account in modeling of the data; however, they can be larger than the statistical uncertainties on the derived model parameters for high signal-to-noise observations (see Section 7.7.5).

4.5.3 For lightcurves

There are no calibration files specifically used in timing analysis. However, there are a few issues that should be considered. As mentioned in Section 4.3.3, if the source lies close to the edge of the detector, it is possible that the aspect dither (or, in the case of Chandra, deliberate dithering) could move the source on and off the detector. For some detectors a very bright source may require a "dead-time correction" due to an instrumental recovery time between registering an event and being available to detect the next event. Although the dead time is automatically calculated, it is often not explicitly

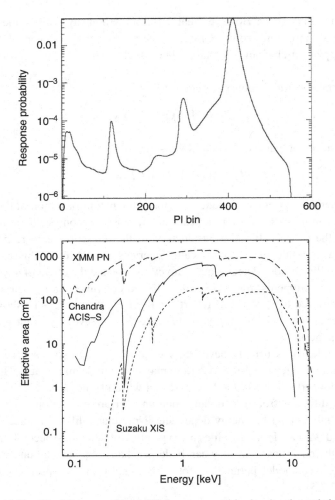

Fig. 4.11   [Top] One row of the ACIS-I response matrix showing the probability
of a 6 keV input X-ray falling in each PI bin. [Bottom] Effective area curves
for the three X-ray missions: XMM–Newton EPIC-pn detector (dashed line),
Chandra ACIS-S (solid line), and Suzaku XIS-BI (dotted line). The effective-
area curves were obtained from PIMMS 4.1 at http://heasarc.gsfc.nasa.gov/Tools/
w3pimms.html

used when making lightcurve plots. When using a software package to do timing analysis, the documentation should be checked to see how dead times are
handled. In extreme cases, the counts from a bright source might be limited by
telemetry saturation as the satellite is unable to transmit the data at the rate it
is accumulated. In general, telemetry saturation is not automatically detected

by software, although it has a characteristic signature: the count rate measured from the entire detector will hover around a peak value, the actual limit. This can be confirmed by checking the documentation available for the mission. Finally, when comparing with observations in other wavelength regimes, it is important to check the time systems used since X-ray event times are generally not barycentered (see Section 4.1.4).

5

Data analysis

RANDALL K. SMITH, KEITH A. ARNAUD, AND
ANETA SIEMIGINOWSKA

5.1 Introduction

This chapter describes some of the data-analysis methods used by X-ray astrophysicists. Any data analysis must begin with careful consideration of the physics underlying the emission before starting to progress through a series of software tools and scripts. After confirming that existing observations could (at least potentially) answer the question at hand, the first step is to determine what observations of the desired source(s) exist. Recent observations are often the best starting point, but even old data are better than nothing. Once some usable data are available the analysis, either spectral, imaging, timing, or some combination of the three, can proceed.

5.2 Low-resolution spectral analysis

5.2.1 General comments

Most recent and current X-ray observations are performed using detectors which provide imaging combined with relatively low spectral resolution. Early missions such as the Einstein Observatory or ROSAT used X-ray mirrors with good imaging capability combined with microchannel plates or position-sensitive proportional counters that had limited spectral sensitivity, typically $R \equiv E/\Delta E \sim 1$–$10$. More recent missions, starting with ASCA, and current missions, such as Chandra and XMM–Newton, use X-ray-sensitive CCDs. These tend to have somewhat higher backgrounds, small pixels, and substantially better spectral resolution, $R \equiv E/\Delta E \sim 10$–$50$, than proportional counters. For comparison, the standard "*UBVRI*" optical filter system is equivalent to $R \sim 4$.

The moderate resolution available with X-ray CCD spectroscopy over the 0.3–10 keV bandpass enables measurements of, for example:

- the column density of absorbing material. This is usually expressed as the equivalent column density of hydrogen, N_H, although at X-ray energies the absorbing material is predominantly helium and heavier elements.
- the electron temperature (T_e) of a hot diffuse plasma, which creates a strong bremsstrahlung continuum (or in the case of an optically thick source, a black-body continuum). Although at low temperatures ($kT_e < 0.5$ keV) radiative recombination of oxygen contributes significantly to the continuum, in general the bremsstrahlung continuum dominates enough line-free regions to allow an unambiguous temperature measurement.
- relative abundances of astrophysically abundant elements such as O, Ne, Mg, Si, S, and Fe in hot thermal plasmas. However, Ne abundances (and to a lesser extent Mg) are often poorly determined because the strong Ne IX and Ne X lines are blended with Fe L-shell transitions (Fe XVII to Fe XXIV).
- the power-law photon index in non-thermal plasmas and detection of any exponential cut-offs if they occur in or near the CCD bandpass.
- Fluorescence emission from Fe Kα at 6.4 keV, created by hard X-rays illuminating cold accretion disks around AGN and X-ray binaries.

However, X-ray CCDs do not have high spectral resolution, a limitation that can lead to difficulties and uncertainties in any analysis. Using dectectors with limited spectral resolution to observe spectra that contain both a strong continuum and emission lines (Figure 5.1) requires a special type of analysis.

The capabilities of X-ray CCDs led to a revolution in X-ray astronomy. Some of the first results from ASCA included measurements of abundances in the core of the Centaurus cluster (Fukazawa *et al.*, 1994), detection of a starburst galaxy with an obscured AGN in NGC 1068 (Ueno *et al.*, 1994), and separation of forward- and reverse-shocked material in the supernova remnant Cas A (Holt *et al.*, 1994).

5.2.2 Spectral fitting

As shown in the previous chapter, the relationship between the source and the observed counts as mediated by the telescope and detector can be described by (Equation 4.10):

$$C(PI) = T \int \text{RMF}(PI, E) \cdot \text{ARF}(E) \cdot S(E) \cdot dE$$
$$\approx T \sum_j R_{ij} A_j S_j \tag{5.1}$$

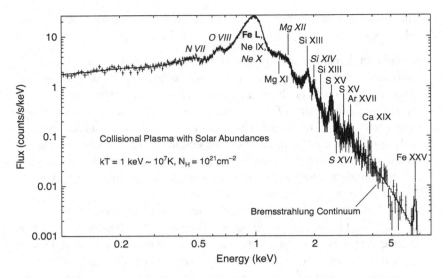

Fig. 5.1   Typical high-quality CCD spectrum of a lightly absorbed 1 keV thermal plasma; note that the spectrum is in detector units of counts, not photons. He-like emission lines are marked in regular type, while H-like lines are in italics and the iron L-shell blend is boldfaced. Emission lines from elements such as O and Si can be resolved, while Ne is blended with the Fe L-shell complex

where $C(PI)$ is the observed counts in detector channel PI, T is the observation time (in seconds), ARF(E) is the energy-dependent effective area of the telescope and detector system (in cm^2), $S(E)$ is the source flux (in photons/cm^2/s/keV), and RMF(PI, E) is the unitless response matrix, or probability of an incoming photon of energy E being observed in channel PI. A convenient approximation shown in Equation 5.1 uses energy bins that are narrow compared to the detector resolution. Each mission provides files containing the response matrix (RMF) R_{ij} and effective area (ARF) A_j, or software to generate them, for every telescope/detector combination. Equation 5.1, a Fredholm integral equation of the first kind, resists easy solution for $S(E)$ by direct inversion of R_{ij}. Statistical and systematic errors in $C(PI)$, calibration uncertainties in R_{ij} and A_j, and the extreme non-linearity of S_j arising from emission and absorption features in the spectrum usually leads to results that are dominated by noise.

The normal approach to low- or moderate-resolution X-ray spectral analysis is to use forward fitting. This begins by specifying a possible model spectrum $S(E)$, such as a power law, bremsstrahlung, or blackbody. The model should be chosen based on the physics of the source. The spectrum is convolved

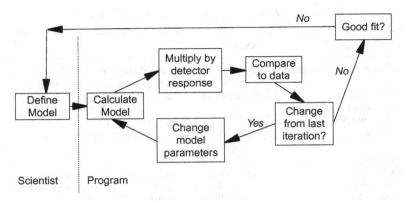

Fig. 5.2   A schematic showing the forward-fitting technique

with the detector response and effective area (as shown in Equation 5.1) and compared to the observed counts $C(PI)$ using some statistical measure (see Section 7.4). The parameters of the initial spectral model are then modified until the best value of the statistic is found. If the best value is still unacceptable, a new spectral model may be selected and the procedure repeated (Figure 5.2). Fortunately, a number of software packages exist to perform forward fitting (see Section 6.3.4). These programs ingest observed spectral data as well as detector responses and effective-area files in their standard formats, include a wide range of spectral models, and use non-linear fitting routines to find the best-fit parameters for each model.

5.2.3 Models (and sources)

The statistician George Box has remarked that "essentially, all models are wrong, but some are useful" (Box and Draper, 1987). It is important to remember that the aim of spectral fitting is to gain physical insight and that all models are likely to be oversimplified in some way.

X-ray spectroscopic models are built up from individual components. These can be thought of as two basic types – additive (an emission component, such as a blackbody or spectral line) or multiplicative (something which modifies the spectrum, such as an edge or absorption line). So, for instance:

$$S(E) = M_1 * M_2 * (A_1 + A_2 + M_3 * A_3) + A_4 \qquad (5.2)$$

where $M_i(E)$ are multiplicative components and $A_i(E)$ are additive.

Multiplicative components just adjust the model by a factor depending on the energy. Models can be modified in more complicated ways by using convolution

Table 5.1   *Physical processes and XSPEC models*

Physical process	XSPEC models[1]
Equilibium collisional plasma	(b)(v)apec, (z)(v)bremss, c6(p)(vm|me)kl, ce(vm|me)kl, equil, (v)meka(l), (v)raymond, smaug
Non-equilibrium collisional plasma	(v)(g)nei, (v)(n)pshock, (v)sedov
Photoionized plasma	absori, redge, swind1, xion, zxipcf
Power-law	bkn2pow, bknpower, cutoffpl, pegpwrlw, plcabs, (z)powerlaw
Blackbody	(z)bbody, bbodyrad
Emission line	(z)gaussian, diskline, kerrdisk, laor(2), lorentz
Compton scattering	bmc, cabs, comp(LS|PS|ST|TT|bb), nthComp, simpl
Accretion disk	disk(bb|ir|m|o|pbb|pn), ezdiskbb, grad, hrefl, kdblur(2), kerr(bb|d|conv), rdblur, sirf, xion
Reflection	(b|p)exr(a|i)v, (i)reflect, refsch
Neutron-star atmosphere	nsa(grav|atmos), nsmax
Cooling flow	(mk)(vm)cflow
Gamma-ray burst	grbm
Pair plasma	nteea
Positronium continuum	posm
Synchrotron	srcut, sresc
Photoelectric absorption	absori, edge, partcov, pcfabs, (v)phabs, pwab, smedge, swind1, tb(var)abs, tbgrain, varabs, wndabs, zvfeabs, zxipcf
Cyclotron absorption	cyclabs
Dust scattering	dust
Reddening	redden, uvred, zdust, zsmdust

The (x) indicates that there are models both with and without the x while (x\|y) means that there are models with either x or y. The (v) indicates that there is an option with variable elemental abundance ratios and (z) indicates an option with redshift as a parameter. The XSPEC model library can be used independently of XSPEC and is also available in the programs Sherpa and ISIS (see Section 6.3.4).

components. These include processes such as smoothing with some function (e.g. velocity broadening), Compton reflection, and detector pileup. If multiple spectra are analyzed at the same time then "mixing" models that link spectra are also useful. An example, is the `project` model that is used in the analysis of clusters of galaxies to go from three-dimensional physical quantities to two-dimensional projected X-ray spectra.[2]

Table 5.1 matches physical processes against names in the XSPEC model library (see Section 6.3.4).

[1] http://heasarc.nasa.gov/docs/xspec/manual/XspecModels.html
[2] http://heasarc.gsfc.nasa.gov/docs/xanadu/xspec/manual/XSmodelProject.html

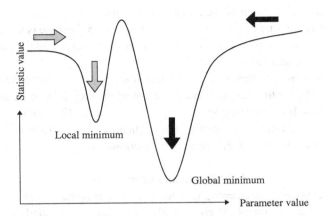

Fig. 5.3 The problem with local minimization. If the fit starts at the right it finds the true minimum but if it starts on the left then it gets stuck in a false minimum

5.2.4 Practical considerations

5.2.4.1 Local minimization

Finding the best fit means minimizing the difference, as quantified by the fit statistic value, between the observed data and that predicted using Equation 5.1. The core problem is how to do the "change model parameters" step in Figure 5.2. Press *et al.* (2007) discuss some of the computationally efficient algorithms available. Most of these choose the new model parameters by using information local to the current parameters. For instance, a simple example is to calculate the derivatives of the statistic with respect to the parameters at the current value of those parameters. The new parameters are chosen by moving downhill. Such algorithms are referred to as local minimization methods and tend to suffer from the problem of getting trapped in local minima. Figure 5.3 illustrates this problem schematically. If the fit process is started at the right place then it will converge to the true minimum but if the wrong starting parameter values are chosen then the fit will end in a local minimum. It is impossible to tell ahead of time the right place to start.

A rule of thumb is that the more complicated the model and the more highly correlated the parameters then the more likely that the algorithm will not find the true minimum. In addition, some specific types of models can introduce problems. For instance, consider the case of a narrow emission line and what happens when the derivative of the fit statistic with respect to the line energy is calculated. The derivative is estimated by changing the line energy by a small amount and recalculating the fit statistic. If the current model line energy is far enough away from that observed then this derivative will be zero so the fit will never change the line energy. One possible solution in this case is to choose an artificially wide line till the fit locks onto the correct energy and then allow the line to become narrower.

If the fit does get stuck in a local minimum it may be possible to stumble over a new, better fit while estimating the confidence region (see Section 7.5). This can then be used as the starting point of a new fit. This crude technique, which can be iterated by estimating the confidence region of the new best fit, is sometimes effective but can be very time-consuming.

5.2.4.2 Global minimization

The problems enumerated above with local minimization have been well known for many years in many different fields so a range of methods have been proposed to ensure that minimizations are truly global. Some of these are based on physical or biological analogs. Simulated annealing allows steps where the fit statistic can increase providing a chance of escaping from local minima. The probability of an increasing statistic step is slowly decreased through the fitting process by analogy to the way that metals cooled slowly anneal into their lowest energy state (e.g. Press *et al.*, 2007). Genetic algorithms are based on the obvious biological analogy and operate on populations of parameter values. Those which provide the best fit "breed" to produce the next generation. These methods require many more function evaluations than local methods so are much slower and are still not guaranteed to find the true minimum.

The most promising and increasingly popular approach is Markov Chain Monte Carlo (MCMC). This bears similarities to simulated annealing but has the major advantage that it can be used simultaneously to search for the best fit and determine the confidence regions. So, although the search for the best fit takes longer than local minimization methods, some of this time is recovered because it is no longer necessary to run a lengthy process to get confidence regions. Figures 5.4 and 5.5 show a simple example use of MCMC to determine a power-law photon index. For an introduction to MCMC by statisticians see Wilks *et al.* (1996) and for astronomical applications see van Dyk *et al.* (2001) (spectral fitting in X-ray astronomy), Verde *et al.* (2003) (determining cosmological parameters from microwave background observations), or Ford (2005) (searching for extrasolar planets).

5.2.4.3 Spectra with few counts

Figure 5.6 shows an example using XSPEC of a common mistake made when fitting X-ray spectra of weak sources. It appears that the spectral-fitting program is unable to match the model to the spectrum with almost all the data points

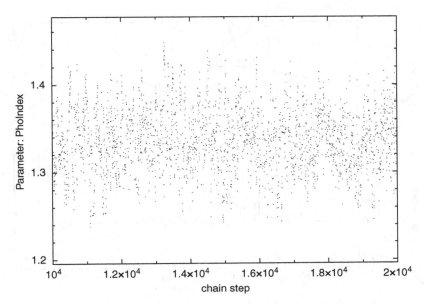

Fig. 5.4 An example use of MCMC. 10 000 steps in a chain showing the power-law index in a fit to an absorbed power-law model

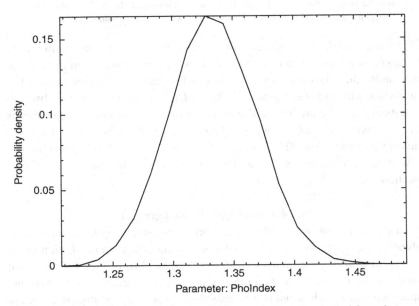

Fig. 5.5 The probability distribution for the power-law index derived from the MCMC chain shown in Figure 5.4

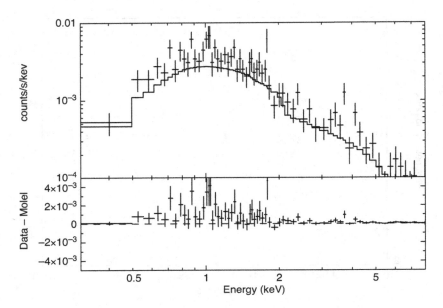

Fig. 5.6   An example showing typical behaviour when $\sum(X_i - M_i)^2/X_i$ is incorrectly used as the fit statistic. The top panel shows the data as points with error bars and the best-fit model as the solid line. The bottom panel shows the differences between the data and the model. The model is systematically below the data

above the model. What has actually happened is that the data points have been grouped together to make a better-looking plot. However, in the statistic calculation the data are ungrouped and many bins have zero or a few counts. The fit statistic which has been used is $\sum(X_i - M_i)^2/X_i$, where X_i are the observed counts and M_i the predicted. The presence of X_i in the denominator means that bins with $X_i < M_i$ are weighted more heavily than those with $X_i > M_i$, driving the model low. The fit-statistic value will also tend to be anomalously low in this case. Since the data are Poisson, the correct statistic to use is C (see Section 7.4.1 and Section 7.7.1).

5.2.4.4 Dealing with background

With the exception of a bright point source observed using Chandra there is always a significant background component to the spectrum in addition to the source of interest. For imaging detectors it is usual to extract a background spectrum from a source-free region. For non-imaging detectors, a background spectrum will be calculated using methods developed by the instrument team. The best method for dealing with the background is simultaneously to fit the source and background spectra with models for the source and background

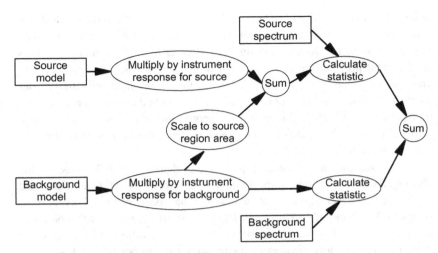

Fig. 5.7 Flowchart (after Broos *et al.*, 2010) showing the procedure for simultaneously fitting source and background

(Figure 5.7). Examples of this approach are found in Broos *et al.* (2010) who use a piece-wise linear model for their Chandra ACIS observations and Siemiginowska *et al.* (2010) who combine an eighth-order polynomial with nine individual Gaussian lines to model Chandra ACIS "blank-sky" data. Background modelling is discussed further in Chapter 8 in the context of analyzing spectra of extended sources that occupy the entire FOV of the instrument.

If a parametrized background model is not available then alternatives can be tried. The simplest case arises if the data have Gaussian errors. Since the difference of two Gaussian distributions is also Gaussian, the S^2 statistic can be used. The observed data are calculated by subtracting the background from the source spectrum, and the variance by adding in quadrature the source and background variances. If the data have Poisson errors the problem is more difficult because the difference of two Poisson distributions does not have a simple, analytic form. The XSPEC program does use a modified version of the C statistic for this case; however, there are issues for very-low-count ($\sim$100) observations so any results should be tested carefully using simulations.

5.2.5 A fully worked example: XMM–Newton observations of Abell 1795

A typical X-ray-analysis project might be to determine the total mass in a cluster of galaxies. Early X-ray observations indicated that galaxy clusters are diffuse X-ray sources due to hot gas in the intracluster medium (ICM) (Kellogg *et al.*, 1973). Soon thereafter, it was suggested that most of the baryonic mass in clusters is in the X-ray emitting ICM (Lea *et al.*, 1973), making X-rays an excellent probe of cluster mass. X-ray measurements of the temperature and flux of the ICM as a function of position, combined with the cluster redshift, determine the cluster luminosity and average density. Then, assuming the cluster is not highly disturbed by a merger, the density and apparent size of the cluster can be combined to make an accurate determination of the total mass (e.g. Lewis *et al.*, 2003).

As an example, we consider a project to calculate the ICM mass of the cluster Abell 1795 within some radius. A search of the archives described in the next chapter shows that this source has been observed by ASCA, BeppoSAX, Chandra, INTEGRAL, ROSAT, Swift, Suzaku, XMM–Newton, and the Einstein Observatory, amongst others. These different telescopes and detectors have widely varying characteristics (see Appendix 4). For example, INTEGRAL is primarily a gamma-ray satellite but since the hot ICM gas does not emit significantly at these energies, these data can be ignored for the moment. A slightly deeper investigation shows that (as of summer 2010) the center of the cluster has been observed 21 times by Chandra (it is used as a calibration source), three times by Swift, and once each by Suzaku and XMM–Newton. Each observation contains a unique Observation Identifier (ObsID); in this case, the longest observation (∼67 ksec) was taken in 2000 by XMM–Newton (ObsID 0097820101). A check of the bibliographic database shows that this observation has been used to date in over 20 different papers. These papers should, of course, be reviewed before beginning any project.

Based on its length, the XMM–Newton observation looks to be a good initial choice. First, check whether there are quick-look products available. If there are, perform a quick, preliminary review of the data to ensure, for example, that (1) the source is in view, (2) that the observing mode is appropriate, and (3) that solar flares did not ruin the observation. Download the data and follow the steps suggested in the last chapter for initial data reduction. If necessary, reprocess the observation to apply updated calibration information. However, reprocessing the data may take some time, and it may not be clear if this is in fact the best observation. Often, performing a quick review of the data as they are initially presented saves time. If the data are then selected for more substantive work, the most recent processing can be applied.

When using imaging X-ray satellites, examining an image of the event file is often a good first step. Figure 5.8 shows an image made from the EPIC-pn CCD data from the XMM–Newton observation of Abell 1795.

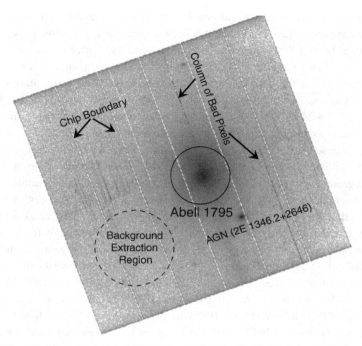

Fig. 5.8 The cluster Abell 1795 observed with the XMM–Newton EPIC-pn detector (ObsID 0097820101), with 3 arcmin radius circle around the cluster core. The FOV also contains a serendipitous observation of a bright AGN, 2E 1346.2+2646

Depending upon the level of processing, this initial image may include bad pixels or columns, as can be seen in Figure 5.8; these will be removed in later processing. At this level, the only issues are whether the data show the source appearing in the expected position (both DS9 and fv allow the position in right ascension and declination to be read directly from the image), whether the source dominates the background, and whether the source is confused with other sources or diffuse emission. In some cases the image will appear strange if the observation was done in an unusual or unexpected mode. Determining the exact mode in use is mission-specific and will be described in the mission documentation.

In the case of Abell 1795, after filtering out most "flare" times (see Section 4.2.1 and 8.2.1) and applying the most recent calibration, the only remaining step before extracting a spectrum is to identify a background region to use with the source. This background can be taken from pre-existing observations, or selected from an apparently source-free region in the observation itself. The latter method was used here, although the cluster is quite bright within the selected region so the exact method is not too important. The source spectrum was taken from a 3 arcmin radius region around the center of the cluster at $(\alpha, \delta) = (13°48'53'', +26°35'19'')$ (J2000). Note that the choice of a 3 arcmin radius means that the final result will be valid for the ICM mass interior to 3 arcmin. The background is from a 4 arcmin radius region centered at $(\alpha, \delta) = (13°49'25'', +26°28'09'')$. All X-ray spectral tools automatically scale any difference in extraction area between the source and background.

The diffuse-cluster X-ray emission arises from optically thin thermal gas in collisional equilibrium and will be absorbed by some amount of foreground material. The gas may have a range of temperatures and emission measures. The spectrum from such a plasma requires detailed models that include both continuum components such as bremsstrahlung radiation and radiative recombination, as well as line emission from radiative transitions between energy levels from all abundant elements excited by electron collisions (see Figure 5.1). Most X-ray astrophysicists use either the APEC model (vapec) based on the AtomDB database of rates (Smith *et al.*, 2001; Foster *et al.*, 2010) or the MEKAL (Mewe *et al.*, 1986; Kaastra, 1992; Liedahl *et al.*, 1995) model (vmekal), although the older Raymond and Smith (1977) model (vraymond) is sometimes used for ease of comparison with earlier papers. A number of absorption models from foreground neutral (or ionized) interstellar medium (ISM) material are used (e.g. phabs); a more recent calculation (tbabs) was done by Wilms *et al.* (2000).

Although clusters have multiple temperature components and possibly multiple layers of internal absorption, the simplest approach begins by assuming a single cluster temperature and foreground absorption component. If an adequate fit can be obtained, it should allow a reasonable estimate of the ICM mass to be made. Using the C statistic in XSPEC, the best-fit parameters are $N_H = (1.08 \pm 0.07) \times 10^{20}/\mathrm{cm}^2$, $kT = 4.75 \pm 0.04$ keV, $z = 0.0627 \pm 0.0001$, and abundances (relative to solar) of 0.46 ± 0.02 (all errors quoted are 90% confidence). Simulations based on this best fit all have values of C below that observed so the model is likely not correct (see Section 7.6). Examination of the residuals (i.e. data − model) shows features at the energies of the Fe L-shell and K-shell lines, which are characteristic of a cluster with multiple temperatures. Nevertheless, the model is adequate when making a simple ICM mass estimate because the total flux is roughly proportional to the density squared but only the square root of the temperature. The total flux in the 0.5–10 keV band is $(5.64 \pm 0.02) \times 10^{-11}$ erg/cm^2/s, corresponding to a luminosity of $(5.34 \pm 0.02) \times 10^{44}$ erg/s at this redshift. All of these values can be easily obtained using simple XSPEC commands (e.g. `fit` or `flux 0.5 10.0`); other spectral packages have similar functions.

The model normalization provides the emission measure of the gas scaled by the distance:

$$\xi \equiv \int n_e n_H dV / (4\pi D_A^2 (1 + z)^2) = (4.33 \pm 0.02) \times 10^{12}/\text{cm}^5 \quad (5.3)$$

where n_e, n_H are the densities of electrons and hydrogen nuclei, and D_A is the angular diameter distance to the source. Using "concordance cosmology"[3] values with the best-fit redshift z, $D_A = 246$ Mpc. Assuming the cluster is spherical, of uniform density, and fully ionized with 10% He (implying $n_e \approx 1.2 n_H$), Equation 5.3 can be rewritten as:

$$n_H = \sqrt{\frac{(1 + z)^2 \xi}{1.2\theta^3 D_A}} = 0.0028 \pm 0.0002/\text{cm}^3 \quad (5.4)$$

using 3 arcmin for the cluster's angular size, θ. The total mass is then simply $m_N n_H V = m_N n_H (4\pi/3)(\theta D_A)^3$, where m_N is the average nucleon mass per hydrogen atom, or 2.12×10^{-24} g/H atom. The final result is an ICM mass of $(3.6 \pm 0.2) \times 10^{12} M_\odot$ within 3 arcmin of the center, in good agreement with more detailed analyses such as that of Ikebe *et al.* (2004; see their Figure 4).

5.3 High-resolution spectral analysis

The spectral resolution of X-ray CCDs is, in general, inadequate to allow purely line-based analysis. This kind of study requires emission or absorption lines to be measured independently of a preconceived model (such as the APEC model used above). Although moderate-resolution data combined with a detailed model can be used to determine model parameters with precision, if the model is fundamentally inapplicable then the fit parameters are meaningless. The power of high-resolution X-ray data to reveal misunderstandings was shown in early analyses of "cooling-flow" clusters, such as Abell 1795. Until the launch of the RGS device on XMM–Newton, these systems were thought to be undergoing dramatic cooling from X-ray temperatures ($T_e > 10^6$ K) down to 10^4 K. Fabian (1994) and many others applied such models to X-ray cluster observations, determining the number of solar masses cooling per year with excellent precision – but, as it turned out, little accuracy. Although the model fits had some problems (Buote, 2000), these could be explained by abundance

---

[3] See http://www.astro.ucla.edu/~wright/CosmoCalc.html (Wright, 2006)

or absorption effects (e.g. White *et al.*, 1991). However, after the launch of XMM–Newton, Peterson *et al.* (2003) used the RGS to measure the spectrum of 14 cooling flow clusters with much higher spectral resolution than CCDs could offer, and immediately showed that the Fe L-shell lines, which were predicted if the gas were truly cooling, were not present. The only explanation was that the cooling of the gas was "cut off" by some unknown process, now thought to be heating from the central AGN.

The exact definition of "high" resolution depends upon the bandpass and the scientific requirements. At a minimum, though, high resolution implies that there are line features which are not significantly affected by blending with other emission lines or the continuum. In an ideal situation, the high-resolution spectrum would allow individual line shapes to be measured directly from the data. At X-ray energies, these two possibilities correspond to $R \approx 300$ and $R \approx 10\,000$, respectively. Above $R = 10\,000$, the line shape from a collisionally excited line in a thermal plasma is dominated by thermal width, limiting the value of increased resolution (although in a photoionized plasma this is not true). Resolutions above 300 are only available at present by using gratings, either transmission for Chandra or reflection for XMM–Newton. In the future, microcalorimeters will provide resolutions up to ≈ 3000.

The initial steps in high-resolution X-ray spectral analysis are identical to those for the low-resolution case: the data must be processed using the latest pipeline and calibration data, checked to ensure the mode is correct, strong flares and bad pixels removed, and so on. The details are, as usual, mission- and detector-specific. To maximize the signal, X-ray grating spectrometers are, in general, slitless. As a result, the spectrum of a point source in a crowded region or inside a region of diffuse emission will be degraded. Spectrometers with a broad dispersion, such as XMM–Newton's RGS, can observe a diffuse source of moderate size ($\theta \sim 1$–$2'$) with only partial degradation. However, Chandra's spectrometers have a narrower dispersion and are severely degraded by a source larger than 10–20''.

Once the initial reduction is complete, the resulting output should include a spectrum with appropriate response and background files. Although, at the highest resolutions, the forward-folding formalism is not required – the spectrum could be divided by the effective area and analyzed in physical units of photons/cm^2/s/Å – many X-ray astronomers stick with familiar methods and continue to use a forward-folding approach. In this case, the response matrix is the line-response function of the spectrometer to a δ-function (see Figure 5.9).

The same warnings about low-count statistics and binning that were mentioned above become even more important when working with high-resolution spectra. Clearly, merging bins in a high-resolution spectrum to increase the

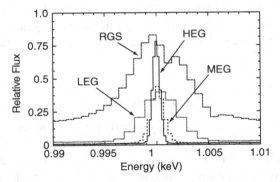

Fig. 5.9 A hypothetical emission line at 1 keV observed with all four major X-ray spectrometers on XMM–Newton and Chandra, showing both the relative flux that would be detected and their line spread function. Although the HEG response is quite narrow and Gaussian in shape, the others – especially the RGS – are quite broad with non-Gaussian features (MEG and LEG are medium- and low-energy gratings, respectively). Thus, the response matrix cannot be ignored in these cases, or the results will dramatically underestimate the power in the lines relative to the continuum emission

number of counts per bin entirely defeats the purpose of obtaining a high-resolution spectrum in the first place. However, the combination of low effective areas on existing satellites, limited observing time, and the brightness of astrophysical X-ray sources ensures that most X-ray-grating spectra will contain many bins with two, one, or even zero counts. Fitting a spectral model using the S^2 statistic will fail since there is no way to estimate the error correctly in the low-count spectral bins. An overestimate of the error in low-count bins leads to fit parameters with strong continua and relatively weak lines, while an underestimate leads to poor fits with weak continua and unphysical line strengths.

There are a few solutions to this problem. Assuming it is not possible to obtain more data, one method is simultaneously to fit the high-resolution spectrum and a CCD spectrum taken using a detector with a larger effective area. This may be straightforward if the observation was done using XMM–Newton, which usually obtains simultaneous CCD and grating spectra. The advantage of fitting both grating and CCD data is that the latter will have many more counts, so the impact of low-count bins to the model fit – especially for the continuum emission – is much lower. However, a Chandra grating observation may have no simultaneous CCD data, and even if data does exist from another observation, the source may have varied between the two observations. If suitable CCD spectra are not available, a good option is to perform the fit using the C statistic described in Section 7.4. Although C does not provide the simple goodness-of-fit measure available when using S^2, the fit values are generally far better than those created using S^2 with some estimation of the errors in the low-count bins.

5.3.1 Line diagnostics

Once initial processing is complete, the next step is often line identification. Published X-ray line lists can be consulted; however, a common tool option is to use AtomDB's WebGUIDE,[4] which creates line lists on demand for any bandpass, using the same atomic database (AtomDB) used by the APEC code. The X-ray band contains features from every ion of every abundant "metal" since the inner 1s K-shell electrons of these elements have binding energies in the range 0.1–10 keV (see Table A1.5). In addition, $n = 2$ (L-shell) transitions[5] generate X-rays for elements from Mg through Ni; the Fe L-shell lines (see Table A1.4) are particularly strong due to the generally high abundance of iron in astrophysical plasmas (see Table A2.3). The energy or wavelength of these transitions depends upon the number of outer electrons present, making them ion-dependent. Of course, the corollary to this is that the vast number of transitions can make identification challenging.

Despite the number of transitions, a few strong lines dominate most diagnostic tests. One of the simplest systems, the Lyman series of lines from hydrogen-like ions (see Table A1.1), behaves as a temperature and ionization-state diagnostic. Figure 5.10 [top] shows the Lyβ/Lyα ratio for a few ions as a function of temperature for a plasma in collisional equilibrium. In addition, a plasma out of equilibrium might show unrealistically large values of Lyβ/Lyα due to charge exchange populating high-n levels, making this a useful consistency check.

Another useful set of diagnostic lines arise from helium-like ions (see Table A1.2). This isosequence contains a "triplet" (actually, a quartet plus a two-photon transition) of strong lines from the $n = 2$ level to the ground state (see Table 5.2 and Figure 5.11). The diagnostic power of these lines in ions from C V to Ni XXVII arises from their wide range of radiative transition rates (typically spanning eight or more orders of magnitude) as well as their different statistical weights in the parent levels. Multiple names are used for these lines; the spectroscopic letters come from Gabriel (1972), while the transition-line name or type (forbidden, intercombination, or resonance) is also used, often

---

[4] http://www.atomdb.org/Webguide/webguide.php
[5] See Section A1.1 for a summary of spectroscopic notation

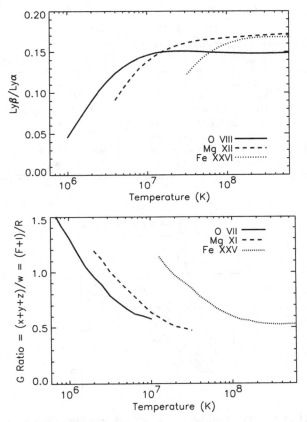

Fig. 5.10 [Top] Lyβ/Lyα ratio for O VIII, Mg XII, and Fe XXVI taken from AtomDB v2.0.0. [Bottom] He-like G ratio for O VII, Mg XI, and Fe XXV from AtomDB v2.0.0

in abbreviation as F, I, and R (or f, i, and r). Gabriel and Jordan (1969) noted that the ratio $R \equiv z/(x + y)$ is a useful density or photon-field diagnostic. The parent level ($^3S_1$ $1s2s$) of the z line is metastable and can be easily excited into the ^3P $1s2p$ level in a sufficiently high-density plasma (or by bright UV radiation, as it is an allowed UV transition for most abundant elements). Gabriel and Jordan (1969) also defined the ratio $G \equiv (x + y + z)/w$ and noted that it is density-*in*dependent, although it is temperature-dependent, as shown in Figure 5.10 [bottom].

In addition to these features, the inner-shell excitation of a Li-like ion creates a line within the He-like triplet known as the "q" line (Gabriel, 1972). The appearance of the q line is usually an indication of an ionizing plasma that is

Table 5.2 *Helium-like isosequence energy levels (n = 1, 2) and strong radiative transitions*

Label	Transition name	Spectroscopic letter
$^1S_0$ $1s^2$	Ground	–
$^3S_1$ $1s2s$	To ground; forbidden (F)	z
$^3P_0$ $1s2p$	To $^3S_1$ $1s2s$ only	No common name
$^3P_1$ $1s2p$	To ground; intercombination (I)	y
$^3P_2$ $1s2p$	To ground; intercombination (I)	x
$^1S_0$ $1s2s$	To ground via two-photon continuum	No line
$^1P_1$ $1s2p$	To ground; resonance (R)	w

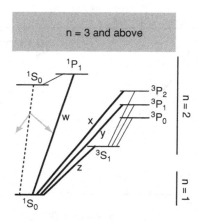

Fig. 5.11 Grotrian energy-level diagram for a Helium-like system, showing the strong $n = 2 \rightarrow 1$ X-ray transitions as thick lines, the $n = 2 \rightarrow 2$ transitions as thin lines, and the two-photon transition as a dashed line

too hot to contain the Li-like ion in equilibrium. This situation might occur in a supernova shock or stellar flare. Many other useful diagnostic lines exist; see Smith (2005) for a recent survey.

5.3.2 Example: line broadening in Abell 3112

Abell 3112 is a nearby, rich X-ray cluster, at a redshift of 0.0746. It has a strong centrally condensed cooling core whose temperature distribution has a lower cut-off at ∼2 keV, probably due to episodic heating by the central AGN. A primary question about galaxy clusters centers on how they are supported: purely by thermal energy, or is turbulence and/or bulk flows significant? One way to address this issue is to measure the width of emission lines in the ICM. Atomic physics sets a fundamental lower limit to any emission line's width, although thermal broadening (see Table A2.2) usually dominates the width of astrophysical X-ray emission lines. Turbulence will also broaden emission lines, while bulk flows may either broaden the line or distort the line shape, depending on their size scale.

Unfortunately, spectra at CCD resolution cannot resolve emission lines sufficiently well to extract useful results. While the Chandra gratings provide high resolution, their relatively small dispersion means that even small clusters are heavily blurred, making clear separation of spatial and spectral structures nearly impossible. However, as noted above, the XMM–Newton RGS gratings have a larger dispersion angle. As a result, when observing moderately extended sources, the RGS line width is $\Delta \lambda = 0.138\,\text{Å} \times \theta$, where θ is the source extent in arc minutes. In the RGS bandpass this resolution significantly exceeds that of CCDs for sources around 1–2 arcmin in size (see Appendix 4).

Figure 5.12 shows a strong line, O VIII Lyα, from an XMM–Newton RGS observation of Abell 3112. Since the RGS is a slitless spectrometer, the line is somewhat broadened by the extended nature of the source. The amount of broadening can be estimated using an image of the source in a bandpass which contains the emission line. The tool `rgsrmfsmooth` from HEAsoft performs this calculation to modify the RGS response matrix based on a source image. Using this method while fitting a Gaussian line and a bremsstrahlung continuum, an upper limit of 430 km/s is obtained for the physical broadening of the line. Alternatively, the source-extent effect can be ignored and the line simply fit assuming a point source. This will obviously increase the apparent broadening, but does provide a robust upper limit if the line is truly narrow and the source relatively small in angular size.

5.4 Imaging analysis

The goals of X-ray imaging analysis are similar to those in other wavebands: identification of sources down to some limit, discrimination between extended and point sources, identification of structure in extended sources, and determining how any of these vary in different spectral bands. Two particular challenges face X-ray astronomers working with images: (1) low count rates and (2) large changes in the PSF over the detector FOV as sources move off-axis.

As described in Section 1.3.2, most X-ray satellites use a Wolter type-I mirror design that is inefficient relative to optical mirrors. The current generation of X-ray satellites therefore use mirrors that are equivalent to less than a half-meter optical telescope. As a result, faint sources may contain as few as $\sim$10 counts in total. Such low count rates demand a careful understanding of detector background, both instrumental and cosmic.

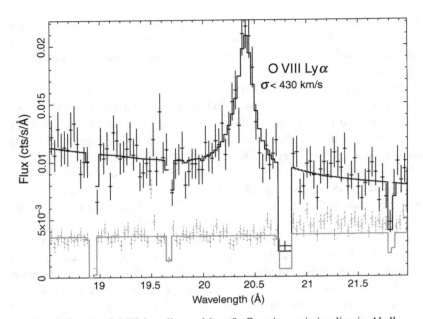

Fig. 5.12   The O VIII Lyα line and best-fit Gaussian emission line in Abell
3112 (in black), together with the background data and fit (in gray). The fit uses
a broadened response to take into account the extent of the cluster, and uses a
doublet of lines with a fixed 2:1 ratio to model the expected contributions from
both components of the O VIII Lyα line. The line is not significantly broadened;
the maximum-velocity broadening(s) of < 430 km/s is a 90% confidence upper
limit

Existing X-ray satellites span a large range in PSF size. The FWHM of the
on-axis PSF is 0.5 arcsec for Chandra and 6 arcsec for XMM–Newton. The
half-power diameter (HPD) for the Suzaku PSF is ≈ 120 arcsec. However,
the Chandra PSF degrades rapidly as sources move more than a few arc minutes
off-axis. This is mirror-specific; the XMM–Newton PSF, for example, degrades
much less. The generally low observed count rates and changing PSF make
source detection to a consistent flux limit both challenging and satellite-specific.
Townsley *et al.* (2006) and Watson *et al.* (2009) include detailed descriptions
using Chandra and XMM–Newton data, respectively.

5.4.1 Source detection

Many source-detection algorithms are available in standard X-ray-data analysis
packages. This section briefly describes a few of the variety of methods used

to detect X-ray sources. The statistical issues raised by source detection are considered in Section 7.7.2.

Sliding-cell (or box) detection methods (e.g. `celldetect` in CIAO, the Chandra data analysis software) determine source and background fluxes simultaneously by convolving the two-dimensional image data with predefined shaped filters. The number of counts in the proposed source and background filter regions are statistically evaluated to determine whether there is an excess of counts above a predefined threshold. The filter size depends on the PSF and can vary across the image. The original sliding-cell algorithm used a box-shaped filter with a larger box for the background than the source. This simple and intuitive method was used in the analysis of the first X-ray images provided by the Einstein Observatory and, later, ROSAT. The sliding-cell method is still used for quick source detection; however, it has some disadvantages including problems in crowded fields and at the detector edge.

A frequently used alternative is to replace the cell filter with wavelet basis functions (e.g. the Mexican Hat) on varying scales (Freeman *et al.*, 2002). This method is computationally very efficient (i.e. fast) and, because the image is usually convolved with wavelet functions on different scales, it is not limited to point sources and locally flat backgrounds. Wavelet detection is more sensitive than cell detection because wavelets can better match the PSF shape (the distribution of photons due to the PSF is not flat over the source filter region). This method nicely resolves close pairs and works well in crowded fields. However, it does tend to resolve an asymmetric source into several individual point sources. In addition, the threshold detection has to be carefully selected for each particular image and tested for false source detection using simulations.

The Voronoi–Tesselation–Percolation algorithm (Ebeling and Wiedenmann 1993; e.g. `vtpdetect` in CIAO) works directly with X-ray events, so the data are not binned and the precise position of each photon is used. It groups photons to compute local background and photon overdensities and makes no assumptions about the observed data except that the background follows the Poisson distribution. The algorithm detects both point sources and diffuse emission (e.g. SNR shells or jets) irrespective of the shape of that emission and its geometry. It is also sensitive to low surface-brightness emission and it is more efficient than the previous two methods in detecting sources. However, blending of close-pair sources can be a problem in crowded fields, so usually the algorithm must be run several times with different detection thresholds.

A different class of detection algorithms uses maximum-likelihood methods (Cruddace *et al.*, 1988; e.g. `emldetect` in XMM–Newton SAS). These methods evaluate the likelihood function for the entire image and find the maximum likelihood for all possible sources given the background. A list of possible source locations is initially constructed (e.g. from a source catalog or by running cell detection). The algorithms perform a simultaneous maximum-likelihood PSF fit to the distribution of source counts in the image. They determine source locations and intensities and estimate uncertainties on these parameters. Some algorithms also give a likelihood of a source being extended. These methods are quite powerful in detecting faint sources in strongly varying background emission and have been used to construct several X-ray catalogs (e.g. ROSAT, XMM–Newton). However, they require efficient optimization algorithms and can be computationally intensive. Consequently, they are often limited to simultaneously fitting a small number of nearby sources rather than the entire field.

5.4.2 Source characterization

Once a potential source has been identified and its flux in one or more bands calculated, the next question is often whether it is a point source or extended. The easiest approach is to calculate the radial profile of the source, and compare this to the mirror PSF, taking into account the position of the source relative to the optical axis of the telescope and the spectrum of the source (e.g. Siemiginowska *et al.*, 2003). All X-ray satellite support centers make available calculations of their satellite's PSF as a function of position and energy, either as simple data files or complete simulators (or both). However, if the source is relatively faint, the PSF is elliptical (see Figure 4.9), or there is a second nearby source, this method may lead to uncertain results. A more complex approach can be done using the CIAO fitting program Sherpa to use the method of forward-fitting described above. This method treats the mirror PSF as the "response" of the telescope and convolves it with a δ-function or a Gaussian to compare against the observed image. If the δ-function model fits the image, then the source is either point-like or unresolved by the telescope. One advantage of this method is that diffuse backgrounds can be included as an additional term in the model.

5.5 Timing analysis

The analysis of time-series data[6] is a common problem across all of astronomy as well other sciences and fields such as electrical engineering. Since X-ray astronomy presents no unique issues and there is a huge literature available, we

---

[6] This section is based heavily on talks given at X-ray astronomy schools by Drs. Michael Nowak and Tod Strohmayer.

limit ourselves here to presenting a few basic results and methods commonly used.

Most X-ray sources are intrinsically variable with timescales ranging from milliseconds to years. Some of this variability may be strictly periodic, as in pulsars, or periodic or quasi-periodic, as in X-ray binaries, but usually there is power across a wide range of frequencies. Some sources burst for short periods of time, others change between states. All of these can provide clues to the physics of the system emitting X-rays. For instance, timing observations of X-ray binaries can provide orbital periods and their evolution as well as the sizes of emitting or occulting regions. If one component of the binary is a neutron star then oscillations at kilohertz frequencies can be used to constrain the neutron-star equation of state.

It is important to be aware of typical timescales that may be imposed upon the data by the instruments or the spacecraft. Among these are readout times of the instrument (e.g. the frame time for CCDs), the orbital period of the spacecraft, and the rotation period of the Earth. Another complicating issue with some instruments is the presence of dead time (Section 4.5.3).

The first step in any analysis is to create a lightcurve. Make sure to use a bin size which is an integer multiple of the time resolution of the instrument. For instance, if the data is from a CCD then the time bin should be an integer multiple of the frame time. However, do not bin the data more than necessary.

Consider a lightcurve, y_k, with bin size Δt and N bins; then the highest frequency for which information can be obtained is the Nyquist frequency, $f_{Nyq} = 1/(2\Delta t)$, and the lowest frequency is $f_{min} = 1/(N\Delta t)$.

5.5.1 Testing for variability

A common first question is whether the observed variation is consistent with noise or whether there is evidence for the source varying. Nandra *et al.* (1997) proposed binning the lightcurve so the counts in each bin can be approximated as Gaussian with error σ_k^2. The excess variance, σ_{rms}^2, is then defined by:

$$\sigma_{rms}^2 = \frac{1}{N\bar{y}^2} \sum_{k=1}^{N} [(y_k - \bar{y})^2 - \sigma_k^2] \tag{5.5}$$

where $\bar{y}$ is the mean of the y_k. The error on the excess variance is $s_D/(\bar{y}^2\sqrt{N})$ where:[7]

$$s_D^2 = \frac{1}{N-1} \sum_{k=1}^{N} ([(y_k - \bar{y})^2 - \sigma_k^2] - \sigma_{rms}^2\bar{y}^2)^2 \tag{5.6}$$

[7] There is a typographical error in this equation in Nandra *et al.* (1997)

An alternative, which does not require binning the data, is to use the Kolmogorov–Smirnov test (e.g. Press *et al.*, 2007) to determine whether the arrival time of events is consistent with a constant rate. Both these methods only estimate whether there is variability; they do not provide any information on the type.

5.5.2 Power spectrum

The most common approach to characterising variability is to derive the power at each frequency. The Fourier transform is used to decompose the lightcurve into a sum of sinusoids with frequencies running from f_{min} to f_{Nyq}. For a lightcurve with N bins and values y_k then the Fourier coefficients are

$$a_j = \sum_k y_k \exp(2\pi i jk/N) \qquad (5.7)$$

where j runs from $-N/2$ to $N/2 - 1$. For computational efficiency N should be a power of 2 to allow the use of fast Fourier transform (FFT) algorithms. Note that some FFT routines will include a normalization factor when calculating a_j in Equation 5.7.

The power spectrum is usually defined as $P_j^{(\text{leahy})} = (2/N_{ph})|a_j|^2$ (Leahy normalization), where N_{ph} is the total number of photons, or as $P_j^{(\text{rms})} = (2/(N_{ph}r)|a_j|^2$ (fractional RMS normalization) where $r = \overline{y}/\Delta t$ is the mean count rate. Since the a_j are complex this removes phase information. In the Leahy normalization the Poisson noise level is 2 and the power scales as the mean count rate, while in the fractional RMS normalization the power is independent of the mean count rate but the Poisson noise level is $2/r$.

Integrating the power spectrum with fractional RMS normalization over frequencies provides a measure of the RMS variability of the lightcurve:

$$\sqrt{\sum_j P_j^{(\text{rms})} f_{min}} = ((\overline{y^2} - \overline{y}^2)/\overline{y}^2)^{1/2} \qquad (5.8)$$

Power spectra are thus often plotted as RMS^2/Hz.

Unlike most other types of data analysis, increasing the length of a lightcurve does not decrease the noise in the power spectrum, it merely reduces f_{min} and distributes the noise over more frequency bins. To reduce the noise, either average over adjacent frequency bins or, equivalently, divide the lightcurve up into sections, calculate the power spectra for each section then average them. This decreases errors by the square root of the number of bins averaged.

Dividing the lightcurve up into sections also allows investigation of whether the power spectrum varies with time. In the study of phenomena such as bursts it is usual to create images with each column being a power spectrum from a section of the lightcurve. Thus, the x-axis is time and the y-axis frequency.

Fourier transforms require that the lightcurve is evenly sampled. If this is not true then a power spectrum can be constructed by least-squares fitting of sinusoids to the data. This is known as the Lomb–Scargle method (Scargle, 1983).

5.5.3 Searching for (quasi-)periodic signals

A simple sinusoidal periodicity will show up as a sharp peak in the power spectrum. The significance of such a peak can be tested if the Leahy normalization is used. In this case, in the limit of large N_{ph}, the power is distributed as χ^2 with two degrees of freedom (Leahy *et al.*, 1983a). If noise reduction has been performed by averaging M frequency bins or lightcurve sections then the noise distribution is modified to χ^2/M with $2M$ degrees of freedom. The standard hypothesis test can now be used to determine the significance of the peak. However, in searching for the periodicity, many frequency bins have been checked so the significance must be reduced by the number of trials, giving a confidence for a detection in frequency bin j of

$$C = 1 - N_{\text{trials}} P_{\chi^2}(M P_j, 2M) \tag{5.9}$$

See also Vaughan *et al.* (1994) for a discussion on upper limits and detection thresholds of periodic signals. For coherent oscillations, the pulsed fraction, f_p, is:

$$f_p = \sqrt{2(P^{(\text{leahy})} - 2)/\bar{y}} \tag{5.10}$$

If the periodicity is not precisely sinusoidal then the power spectrum will have multiple peaks with harmonics at integer multiples of the base frequency. Further, if there are instrumental periodicities then not only will these appear in the power spectrum but there may be beats showing up at the difference between the source and instrumental periodicities (and their harmonics).

While some periodicities detected in X-ray-astronomy data are sharp, many have a significant width. This width may arise because the periodicity is only on for part of the lightcurve or the oscillator may not be coherent, changing in phase or frequency. Coherence is characterized by $Q = f/f_{\text{FWHM}}$. Quasi-periodic oscillations (QPOs) may be hard to find and will require searching using multiple binning scales. In general, sensitivity is maximized when the frequency resolution of the power spectrum approximately matches the width of the QPO. Of course, when estimating the significance of the signal, the number of different trial searches must be taken into account, or else the significance will be overestimated.

Highly non-sinusoidal periodicities can be hard to detect in the power spectrum because their signal is spread across so many separate harmonics. In this case it may be better to use epoch folding or the Rayleigh test (Leahy *et al.*, 1983b). In both these methods a test period is chosen and its pulse profile used to calculate a statistic (Davies, 1990). The test period is varied over a range and the best statistic value found.

After performing any of these methods and finding a periodicity, the lightcurve can be folded on the period to produce a pulse phase plot.

5.5.4 Variability estimates when writing proposals

Useful approximate formulae for determining detectable variability are as follows. For broad-band variability:

$$\text{RMS}^2 = 2n\sqrt{\Delta f}/\sqrt{Rate^2 \times Time} \qquad (5.11)$$

where n is the statistical significance required (in sigma), Δf is the frequency bandwidth of the signal, *Rate* is the expected count rate and *Time* the proposed observation time. For coherent pulsations

$$f_\text{p} = 4n/(Rate \times Time) \qquad (5.12)$$

5.5.5 Bayesian methods

Bayesian analyses have led to a couple of methods which have been applied in X-ray astronomy. Gregory and Loredo (1992) presented a method for detecting a periodic signal of unknown shape and period. One advantage of this algorithm is that it works well even when there are gaps in the data. A disadvantage is that it does not provide a measure of the signal significance but only odds ratios between alternative models.

Scargle (1998) introduced a method known as Bayesian Blocks. This is an algorithm that identifies time segments in the raw event data (so no binning is required) over which there is no statistically significant variation. This provides an optimal binning scheme, which can find variability potentially down to the time resolution of the event data.

A class of recently developed methods is based on modeling the lightcurve assuming that variability is a result of a mixture of stochastic processes (Uttley *et al.*, 2005; Kelly *et al.*, 2009, 2010). These methods are independent of the binning used and can be applied to lightcurves that have gaps or uneven sampling. They can be used to place limits on characteristic variability timescales and, potentially, link the variability to physical processes.

6

Archives, surveys, catalogs, and software

KEITH ARNAUD

6.1 Archives

Unlike other branches of astronomy that have proprietary telescopes, X-ray astronomy is of necessity done using satellites and suborbital rockets. These are funded by national governments, which typically insist that all data be made public within a reasonable time. As a result, practically all X-ray-astronomy observations are available on-line; the challenge is to find the right data. Fortunately, X-ray-astronomy data are concentrated in a small number of archives and it is usually clear which website to try.

The best way to think of an archive is as a collection of tables, some of which have data sets attached. Some of the tables are simple catalogs, e.g. a list of stars with positions, spectral types, and fluxes. Other tables come with considerable data attached, e.g. the observation catalog for some mission that lists pointing position, exposure time and so forth but also provides links to all the publically available data for the observation. These data may include basic event files, cleaned event files, auxiliary information such as housekeeping and orbit files, and product files such as spectra, images, and lightcurves. The archive Internet interface will generally allow the astronomer to choose which categories of data to download for the selected observation.

As an example, consider finding Chandra observations of the Perseus cluster. Its name can be entered in the Chandra data archive search page and the NED[1] or SIMBAD[2] servers used to translate the name into a position. This second step is necessary because the Perseus cluster goes by several names and there may be observations listed under one of its many aliases. Figure 6.1 shows the result of the search. Each row of the table presents information about one observation which satisfied the search criterion. The row or rows which look

[1] http://nedwww.ipac.caltech.edu
[2] http://simbad.u-strasbg.fr/simbad

Select	Row	Seq Num	Obs ID	Instrument	Grating	Appr Exp	Exposure	Target Name	PI Name
☐	1	700005	333	ACIS-S	HETG	50.0	27.41	NGC 1275	Canizares
☐	2	700201	428	ACIS-S	HETG	25.0	25.03	NGC 1275	Canizares
☐	3	800010	502	ACIS-I	NONE	5.0	5.38	A426	Fabian
☐	4	800011	503	ACIS-S	NONE	10.0	9.13	A426	Fabian
☐	5	800011	1513	ACIS-S	NONE	25.468	25.06	A426	Fabian
☐	6	800209	3209	ACIS-S	NONE	97.0	97.04	ABELL 426	Fabian
☐	7	800209	3404	ACIS-S	NONE	5.0	5.86	ABELL 426	Fabian
☐	8	800209	4289	ACIS-S	NONE	96.5	96.68	ABELL 426	Fabian
☐	9	800397	4946	ACIS-S	NONE	29.0	23.98	Abell 426	Fabian
☐	10	800397	4947	ACIS-S	NONE	30.0	30.18	Abell 426	Fabian

Fig. 6.1 The result of searching the Chandra archive for observations of the Perseus cluster. Note that the target name, which was selected by the proposer, varies from row to row although all these observations are of the same region on the sky

most useful can be selected and the primary and/or secondary data products downloaded.

6.1.1 HEASARC

NASA's repository for X-ray (plus gamma-ray and now cosmic microwave background) data is the HEASARC.[3] A collaboration between the GSFC and the Harvard–Smithsonian Center for Astrophysics (CfA), the HEASARC provides access to public data for all currently operating X-ray missions and all major past missions for which data recovery was possible. It includes both US and non-US missions. The HEASARC serves many other tables including major astronomical catalogs and observing logs for some missions operating in other wavebands (such as the Hubble and Spitzer space telescopes). Some of the data sets and catalogs available through the HEASARC are actually hosted elsewhere but the HEASARC provides seamless access. This is the case for both the Chandra and XMM–Newton data sets.

The HEASARC data archives can be accessed using several methods.[4] There is a simple Google-like search box as well as more formal interfaces, which provide guidance in addition to advanced options. One of these options is to cross-correlate HEASARC tables. The entire archive is also available directly

---

[3] http://heasarc.gsfc.nasa.gov
[4] http://heasarc.gsfc.nasa.gov/docs/archive.html

using http or ftp. If many data sets are required, the HEASARC command line batch interface[5] can be used.

Hera[6] is a software service provided by the HEASARC. Instead of downloading a data set and appropriate software to a home machine, the analysis can be run on a server at the HEASARC. This eliminates download time and bandwidth concerns as well as the sometimes difficult task of installing new software packages. It is particularly useful for exploratory analysis to test an idea or to decide whether it is worth investing the time and effort to do a more complete analysis on a local machine. Hera is run through either the fv program or, with some restrictions, a web browser.

6.1.2 Chandra

The CXC[7] is responsible for all aspects of Chandra operations including hosting the current Chandra archive. Although the Chandra archive can be reached through the HEASARC interface, the CXC's WebChaSer[8] provides access which is more focused and includes additional information. Pressing the "View Observation Information" button after searching, then selecting an observation from those returned, gives options for additional information. Particularly useful are the publications link, which gives a list of papers published using data from the observation, the V&V link, which provides the verification and validation report from the CXC staff member who checked the data for anomalies, and the details link which includes the total number of counts observed.

To help decide whether an observation is of interest, the Chandra Fast Image[9] option can be used to take a quick look at an image for a given object or position.

6.1.3 XMM–Newton

The XMM–Newton SOC[10] at Vilspa is responsible for scientific operations of XMM–Newton. This includes managing the archive, which is served through the XSA interface. XSA[11] is a web application that runs as a Java applet. It uses the "shopping basket" model for selecting observations then scheduling them for download. Like WebChaSer for Chandra, XSA provides additional information not available through the standard HEASARC interfaces. These details include bibliographic information and low-level housekeeping logs. For the expert user, XSA supports queries in SQL.

---

[5] http://heasarc.gsfc.nasa.gov/W3Browse/w3batchinfo.html
[6] http://heasarc.gsfc.nasa.gov/hera   [7] http://cxc.harvard.edu
[8] http://cda.harvard.edu/chaser   [9] http://cda.harvard.edu/pop
[10] http://xmm.esac.esa.int   [11] http://xmm.esac.esa.int/xsa

The Browsing interface for RGS Data (BiRD)[12] displays quick-look spectra and images from the RGS, and can also combine multiple spectra, optionally taking into account calibration effects, Galactic absorption, and redshift.

6.1.4 Other archives and mirrors

Several organizations maintain their own archives or mirrors, one of which may work better when downloading from locations outside the USA. These sites include : LEDAS[13] at Leicester University in the UK for Chandra, ASCA, ROSAT, and Ginga; DARTS[14] at ISAS in Japan contains complete archives for Suzaku, ASCA, Ginga, and Tenma; ASDC[15] at ESRIN in Italy includes data from Chandra, XMM–Newton, Swift, BeppoSAX and ROSAT.

6.1.5 Virtual observatory (VO)

All X-ray astronomy archives make their data available to VO tools. Thus, DataScope[16] can be used to find all X-ray data associated with a sky position or astronomical object. Although X-ray astronomers are not yet heavy users of VO tools, their use will likely grow with time and all major archives are committed to supporting the VO.

6.2 Surveys and catalogs

This section describes the most important X-ray surveys and catalogs. Many of these are fertile sources of targets for extended study by current X-ray observatories or follow-up in other wavebands. Individual catalogs are identified by their HEASARC names, which are given in italic script. For more information about a catalog find it in the list at http://heasarc.gsfc.nasa.gov/cgi-bin/W3Browse/w3catindex.pl then click on the name.

6.2.1 ROSAT All-Sky Survey (RASS)

The ROSAT, launched in 1990, was a German/US/UK collaboration. Its first six months in orbit were spent performing an all-sky survey in the

---

[12] http://xmm.esac.esa.int/BiRD/index.html   [13] http://ledas-www.star.le.ac.uk
[14] http://www.darts.isas.jaxa.jp   [15] http://www.asdc.asi.it
[16] http://heasarc.gsfc.nasa.gov/cgi-bin/vo/datascope/init.pl

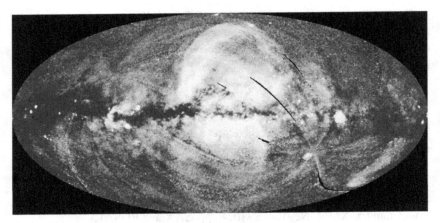

Fig. 6.2   All-sky (Galactic coordinates in Aitoff projection) map of the RASS in the energy range 0.5–0.9 keV

0.1–2.5 keV band, now generally known as the ROSAT All-Sky Survey or RASS. This produced catalogs of individual sources (discussed below) as well as maps of the diffuse Galactic emission in six energy bands with about 12 arcmin angular resolution (Snowden *et al.*, 1997). These maps are available through SkyView.[17] An example all-sky map is shown in Figure 6.2.

6.2.2 ROSAT point-source catalogs

The RASS Bright Source Catalogue (*rassbsc*; Voges *et al.*, 1999) includes all point sources with count rates above 0.05 counts/sec in the 0.1–2.4 keV ROSAT PSPC band. The flux limit depends on the assumed spectrum of the sources but is approximately 5×10^{-13} erg/cm^2/s. This catalog was extended to all sources with detection probability better than 0.999 and at least six source photons to give the $\approx$ 106 000 objects in the Faint Source Catalogue (*rassfsc*). These ROSAT sources have been cross-correlated with catalogs at other wavelengths such as those from the Two Micron All-Sky Survey (*rass2mass*) and the Sloan Digital Sky Survey (*rassdssagn* and *rassdsstar*).

Following its initial six-month survey, ROSAT spent the next eight years performing pointed observations. Serendipitous source catalogs based on the standard processing are available through the HEASARC as *rospspc* and *roshri* for PSPC and HRI observations, respectively. Independently generated catalogs for the PSPC and HRI are also available as *wgacat* and *bmwhricat*.

---

[17] http://skyview.gsfc.nasa.gov

6.2.3 ROSAT cluster catalogs

The RASS is the best source for complete samples of X-ray-selected clusters of galaxies and there are several catalogs based on optical follow-up of extended sources. The main catalogs are *noras* in the north and *reflex* in the south. A subset of the brightest clusters in the north, the extended Brightest Cluster Survey (*rassebcs*), has provided many targets for Chandra and XMM–Newton.

The pointed phase observations have also been used to produce a catalog of serendipitously observed high-redshift clusters. This 400-square-degree catalog (*ros400gcls*) provides a cosmologically important sample of clusters at redshifts beyond 0.3.

6.2.4 Chandra source catalogs

The Chandra Source Catalog[18] (Evans *et al.*, 2010) is ultimately intended to be the definitive list of all sources detected using Chandra. The initial release includes just point and small (< 30 arcsec) sources from the first eight years of observations. Each source in the catalog has an event file, image, lightcurve, spectrum, and response, allowing quick scientific analysis.

A large collaboration headed by members of the CXC is carrying out a multi-wavelength identification program on these sources. Extragalactic sources are being followed up by ChaMP[19] (*champpsc*) and those in the Galaxy by ChaM-Plane[20] (*champlane* and *champlanex*).

There are two large catalogs of serendipitous sources created independently of the CXC. Xassist[21] (*cxoxassist*) is an automated processing of all public Chandra observations to produce source lists and products such as spectra and responses. The Brera Multi-scale Wavelet Chandra Catalog[22] (*bmwchancat*) was created using a wavelet source-finding algorithm on all ACIS-I observations with exposures longer than 10 ksec in the first three years of Chandra operations. This catalog is more sensitive for extended sources than the CSC or Xassist.

6.2.5 Chandra extragalactic surveys

A number of teams have used Chandra to perform deep surveys over regions of the sky. The flagship programs are the Chandra Deep Field North (2 Msec; CDFN or GOODS-North) and Chandra Deep Field South (2 Msec; CDFS or

---

[18] http://cxc.harvard.edu/csc   [19] http://hea-www.harvard.edu/CHAMP
[20] http://hea-www.harvard.edu/ChaMPlane   [21] http://xassist.pha.jhu.edu/zope/xassist
[22] http://www.brera.inaf.it/BMC/bmc_home.html

Table 6.1   *Chandra extragalactic surveys*

Name	Sky area (arcmin^2)	Exposure (ksec)	HEASARC catalog name
CDFN	448	2000	*chandfn2ms*
CDFS	436	2000	*chandfs2ms*
CDFS-Extended	900	250	*chanextdfs*
CLASXS	1440	40	*clasxs*
COSMOS	1800	160	*ccosmoscat*
Elais	594	75	*elaiscxo*
Groth strip	2400	200	*gwsstrpcxo*
Lockman hole	2160	70	*clans*
Spices/Lynx	296	185	*spicescxo*
SSA22	330	392	*ssa22cxo*
XBOOTES	32400	5	*xbootes*

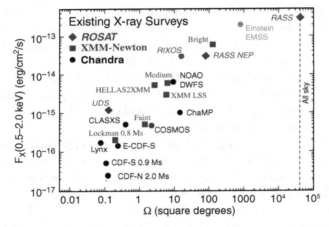

Fig. 6.3   The flux and sky coverage of selected extragalactic surveys. Based on Figure 1 from Brandt and Hasinger (2005)

GOODS-South). The CDFS program also has four flanking fields observed to 250 ksec each. At the time of writing, additional observations of the CDFS are extending its depth to 4 Msec. Other surveys have been performed either on targets with extensive existing data and/or in combination with other observatories. Table 6.1 lists these surveys with their sky coverage and sensitivity. Extragalactic surveys from the first five years of Chandra were reviewed by Brandt and Hasinger (2005), from whom we have taken Figure 6.3, which shows the depth and sky coverage of some of these surveys along with comparable surveys using other observatories.

Table 6.2   *Chandra young clusters and star-formation regions*

Name	ksec	Catalog	Name	ksec	Catalog
Arches/Quint.	100	*arcquincxo*	Cep B/OB3	30	*cepbob3cxo*
Chamaeleon	66	*chainthcxo*	Orion	838	*coup*
Cygnus OB2	147	*cygob2cxo2*	H Persei	41	*hperseicxo*
IC 1396N	30	*ic1396ncxo*	IC 348	53	*ic348cxo*
NGC 2264	100	*ngc2264cx2*	NGC 2362	98	*ngc2362cxo*
NGC 2516	135	*ngc2516cxo*	NGC 6334	80	*ngc6334cxo*
NGC 6357	38	*ngc6357cxo*	NGC 6530	60	*ngc6530cxo*
OMC-2/3	100	*omc2p3cxo*	RCW 108	90	*rcw108cxo*
RCW 38	97	*rcw38cxo*	RCW 49	40	*rcw49cxo*
Rho Ophiuchi	197	*rhoophcxo*	Rosette	60	*rosettecxo*
Trifid (M20)	58	*trifidcxo*	Westerlund 1	60	*wd1cxo*
M17	40	*m17cxo*	NGC 1333	38	*ngc1333cxo*
NGC 2024	76	*ngc2024cxo*	NGC 2244	94	*ngc2244cxo*

6.2.6 Chandra Galactic center survey

Chandra's superb spatial resolution and bandpass up to 10 keV enable it to see through the intervening matter to image the Galactic center region. Wang *et al.* (2002) surveyed a $2° \times 0.8°$ region centered on Galactic longitude $(l) = -0.1°$, latitude $(b) = 0.0°$ with 30 separate 12 ksec Chandra observations. This region has been observed repeatedly and Muno *et al.* (2009) present a catalog of sources based on the accumulated 2.25 msec of Chandra observations up to August 2007.

6.2.7 Chandra catalog of types of object

Chandra has performed deep observations of many young clusters and star-formation-regions, accumulating catalogs of thousands of young and pre-main-sequence stars. Table 6.2 lists the current catalogs. In addition, the ANCHORS[23] project is generating an archive of Chandra star-formation-region observations, which have been processed and analyzed in a uniform manner.

A catalog of supernova remnants observed using the ACIS instrument is available through ChandraSNR.[24] This includes images, spectra, and derived quantities. It is intended as an aid to proposal writing, not a substitute for the researcher's own analysis.

[23] http://cxc.harvard.edu/ANCHORS
[24] http://hea-www.harvard.edu/ChandraSNR/index.html

Table 6.3   *Chandra nearby galaxies*

Name	ksec	Catalog	Name	ksec	Catalog
M31 bulge	5	*m31cxo2*	M31 core	4	*m31cxoxray*
M33	100	*m33chase*	M81	50	*m81cxo*
M87	154	*m87cxo*	NGC 4472	40	*ngc4472cxo*
NGC 4636	193	*ngc4636cxo*	NGC 4649	37	*ngc4636cxo*

Several nearby galaxies have been targets of Chandra large or very large projects with total exposure times in the hundreds of kiloseconds. These observations of M33, M82, and M101 are rich resources.

A number of catalogs are available listing point sources detected in nearby galaxies. *channsgpsc* is based on a total exposure of 869 ksec of 11 nearby, nearly face-on spiral galaxies. The catalog *chpngptsrc* includes 1441 point sources detected in 32 nearby spiral and elliptical galaxies. Table 6.3 lists catalogs of sources in individual galaxies.

XJET[25] provides a catalog of jets found in Chandra observations of radio galaxies and clusters. Radio and X-ray superimposed images are available and, in some cases, FITS images and event files.

6.2.8 Chandra grating observation catalogs

The Chandra HETG team at MIT have created an archive of processed grating spectra and associated response files. The TGCat archive[26] can be used to search for observations, see quick-look plots, check for papers based on the data, and download spectra and responses.

The X-Atlas database[27] (Westbrook *et al.*, 2008) comprises HETG spectra, initially of stars, but expanded to other point sources. It includes the results of spectral fitting although these may be meaningless for targets which are not stars.

The HotGAS database[28] contains HETG spectra of type I and type II AGN which became public prior to July 2004.

6.2.9 XMM–Newton source catalogs

The XMM–Newton serendipitous source catalog[29] (*xmmssc*; Watson *et al.*, 2009) from the XMM–Newton Survey Science Consortium lists sources

---

[25] http://hea-www.harvard.edu/XJET   [26] http://tgcat.mit.edu
[27] http://cxc.harvard.edu/XATLAS   [28] http://hotgas.pha.jhu.edu
[29] http://xmmssc-www.star.le.ac.uk/Catalogue

detected in all publically available XMM–Newton observations. There is a coordinated optical imaging and spectroscopic follow-up campaign. The Internet interface FLIX[30] provides an estimate of the flux or an upper limit for any position on the sky which has been observed.

An independent search for serendipitous sources in XMM–Newton public data is provided by Xassist[21] (*xmmxassist*).

The XMM–Newton detectors operate while the satellite slews between targets and the EPIC-pn has a short enough frame time that objects passing through the FOV can be imaged as point sources. The 0.2–12 keV detection limit is 1.2×10^{-12} erg/cm²/s with a coverage of 50% of the sky expected by the end of the mission. In the ROSAT PSPC energy band the slew survey sensitivity is comparable to that of the ROSAT All-Sky Survey. The point-source catalog[31] is updated approximately annually.

The XCS survey[32] is a search through the entire XMM–Newton public archive to build up a sample of serendipitously observed clusters of galaxies (Romer *et al.*, 2001).

6.2.10 XMM–Newton extragalactic surveys

Although the XMM–Newton spatial response is worse than Chandra it is still good enough for useful deep surveys. Many fields have now been covered by both observatories, so Chandra data can be used to eliminate source confusion and determine precise positions; after which XMM–Newton data can be used to investigate X-ray spectra of the detected sources. Table 6.4 lists some of these XMM–Newton surveys with their sky coverage and exposure. Sky area and exposure times are approximate since in some surveys effective exposure time varies across the survey region.

6.2.11 XMM–Newton catalogs of types of object

XMM–Newton has observed a number of young clusters and star-formation regions leading to the catalogs of sources listed in Table 6.5.

XMM–Newton has performed surveys of the nearby galaxies M31 (*m31xmmxray* and *m31xmm2*), M33 (*m33xmmxray* and *m33xmm2*), and NGC 253 (*ngc253xmm*).

---

[30] http://www.ledas.ac.uk/flix/flix.html
[31] http://xmm.esac.esa.int/external/xmm_products/slew_survey
[32] http://xcs-home.org

Table 6.4   *XMM–Newton extragalactic surveys*

Name	Sky area (arcmin^2)	Exposure (ksec)	HEASARC catalog name
Elais S1	2160	90	*elaiss1xmm*
Lockman hole	700	700	*lockmanxmm*
Marano	2160	120	*maranoxmm*
13-hr field	650	120	*ros13hrxmm*
SA 57	720	67	*sa57xmm*
Subaru field	4100	35	*sxdscat*
COSMOS	7670	40	*xmmcosmos*
XMM-LSS	14300	15	*xmmlss*
Groth strip	1040	50	*xmmcfrscat*

Table 6.5   *XMM–Newton young clusters and star-formation regions*

Name	ksec	Catalog	Name	ksec	Catalog
Carina OB1	150	*carinaxmm*	Cep A	44	*cepaxmm*
Lupus	22	*lupus3xmm*	NGC 2264	42	*ngc2264xmm*
NGC 2516	147	*cygob2cxo2*	NGC 2516	106	*ngc2516xmm*
NGC 6231	180	*ngc6231xmm*	NGC 6530	20	*ngc6530xmm*
Sigma Ori	43	*sigorixmm*	Taurus	35	*taurusxmm*
Upper Sco	50	*upprscoxmm*	NGC 2547	49	*ngc2547xmm*

6.2.12 Swift BAT survey

The Swift Burst Alert Telescope (BAT) has a two-steradian FOV so, while watching for gamma-ray bursts and pointing at specific targets, it is building up the most sensitive survey of the sky in the 14–195 keV hard X-ray band.

The current catalog[33] is based on the first 22 months of the mission providing 461 sources to a detection sensitivity of 2.2×10^{-11} erg/cm^2/s in the BAT band. The Swift BAT team also produce lightcurves of selected sources.[34]

6.2.13 RXTE ASM catalog

The RXTE All-Sky Monitor (ASM)[35] covers 80% of the sky every 90-minute orbit of RXTE with an energy range of 2–10 keV, a spatial resolution of $3' \times 15'$, and a sensitivity sufficient to monitor 350 sources, primarily bright

[33] http://swift.gsfc.nasa.gov/docs/swift/results/bs22mon
[34] http://swift.gsfc.nasa.gov/docs/swift/results/transients
[35] http://heasarc.gsfc.nasa.gov/docs/xte/asm_products.html

X-ray binaries and active galactic nuclei. These sources, not all of which are active all the time, are listed in the ASM source catalog and their current status shown in the ASM X-ray weather map.

6.2.14 MAXI catalog

The MAXI experiment on the Japanese section of the International Space Station operates in the energy range 0.5–30 keV and is expected to monitor over 1000 sources. Lightcurves are available through the data products archive[36] which is updated daily.

6.3 Software

The plethora and variety of software available for X-ray astronomy can be confusing. It is useful to distinguish between two basic classes of software tools: mission-dependent and mission-independent. Every individual mission team will, as part of their basic operations, write software tools to perform tasks specific to the mission. These may be used primarily in the standard processing or by individual researchers, who will have to install the tools as part of their data analysis. However, there are other tasks which are common across many missions such as basic filtering of events by region, time, or other criteria and the creation of images, spectra, and lightcurves. In an ideal world, there would be one system to download and install, which would include all the tools, both mission-dependent and independent, required to analyze X-ray astronomy data. In practice, the situation is more complicated. Depending on the mission under consideration and the analysis to be performed, several packages may need to be installed. The following sections describe the major systems.

The HEASARC maintains a web page[37] tracking latest updates of major astronomical packages. This is a useful resource to see what is available and to help keep up-to-date. Note that while there are X-ray-astronomy packages within the IRAF system (the software package from the National Optical Astronomy Observatory) these are not standard and in many cases obsolete.

This section describes the major packages then considers specific types of multi-mission software.

---

[36] http://maxi.riken.jp/top
[37] http://heasarc.nasa.gov/docs/heasarc/astro-update

6.3.1 Packages

6.3.1.1 HEAsoft

The HEAsoft package,[38] sometimes also referred to as FTOOLS, is a large software collection from the HEASARC. It includes mission-dependent tools for a number of satellites, notably Swift, Suzaku, Fermi, RXTE, and ROSAT, as well as a suite of mission-independent programs. It runs on all common operating systems and the support team will offer help to anyone trying to build on a new variant of the Linux operating system. Both source and binary distributions are available but it is generally better to download the source version and build on a local machine – it only takes a couple of hours and eliminates possible conflicts between system library versions. HEAsoft is required for the analysis of all the missions listed earlier as well as many older missions. It is also useful but not essential for analyzing data from XMM–Newton and Chandra.

6.3.1.2 CIAO

The standard software package for Chandra, CIAO,[39] is produced by the CXC. It is required for Chandra data analysis since it contains the complete suite of mission-dependent tools. However, many of the tools can be used on data from other missions and CIAO includes a suite of powerful higher-level programs. The basic distribution is binary with the major operating systems supported. A source distribution is available, but is for the expert. CIAO is very extensively documented on the CXC website with many useful threads describing how to perform common analysis tasks.[40]

6.3.1.3 XMM–Newton SAS

The XMM–Newton SOC is responsible for the SAS which includes the mission-dependent tools required to analyze XMM–Newton observations. These come in command-line or graphical user interface (GUI) versions. However, SAS does not include higher-level software so it must be used in concert with HEAsoft (or, less commonly, CIAO). The SAS is only available in a binary distribution on a subset of specific systems.[41] To learn how to use SAS read the User's Guide[42] and the XMM–Newton ABC Guide.[43]

6.3.1.4 IDL

There are several packages available for X-ray analysis using the commercial IDL system. The Chandra ACIS hardware team have released the

[38] http://heasarc.gsfc.nasa.gov/docs/software/lheasoft   [39] http://cxc.harvard.edu/ciao
[40] http://cxc.harvard.edu/ciao/threads/index.html
[41] http://xmm.esac.esa.int/sas/current/download
[42] http://xmm.esac.esa.int/external/xmm_user_support/documentation/sas_usg/USG
[43] http://heasarc.gsfc.nasa.gov/docs/xmm/abc/abc.html

TARA[44] package of tools for visualization and analysis including the ACIS Extract program, which can be used to automate analysis of Chandra fields containing many point sources. A large library of IDL procedures for astronomy are maintained at GSFC as the IDL Astronomy User's Library.[45]

6.3.2 Software to manipulate event files

Several programs are available to filter X-ray event files and create the products necessary for higher-level data analysis. The HEAsoft program xselect can be used to filter event files in all the ways described in Chapter 4. It uses a mission database file to hold information on many instruments and can be used automatically to make appropriate calibration products. xselect itself runs a lower-level program called extractor, which can also be used stand-alone.

In CIAO, filtering of event files is performed using dmcopy, which has a wide range of options[46] and the extraction of products by dmextract. The latter uses a defaults file which is compatible with the xselect mission database file.

For XMM–Newton data, SAS includes a program evselect which performs the same tasks as xselect and dmcopy/dmextract. In addition, SAS also has a GUI, xmmselect, which can be used to drive evselect and display the results.

6.3.3 Imaging-analysis software

Much of the imaging analysis in X-ray astronomy is similar to that performed in the optical so, with some caveats, standard astronomy software can be used. The overwhelmingly most popular image-analysis tool is DS9[47] from SAO. Of particular use in X-ray astronomy is its ability to read event files, which are binned up into images, and to construct regions and write them in the formats used by HEAsoft, CIAO, and SAS.

The HEAsoft package includes two programs that are used for image analysis. The FITS viewer, fv,[48] can create images from any two columns in an event file then display either in pow or DS9. The command-line driven ximage[49] program performs a number of X-ray imaging standard tasks. Particularly

---

[44] http://www.astro.psu.edu/xray/docs/TARA
[45] http://idlastro.gsfc.nasa.gov/homepage.html
[46] http://asc.harvard.edu/ciao/ahelp/dmfiltering.html
[47] http://hea-www.harvard.edu/RD/ds9   [48] http://heasarc.gsfe.nasa.gov/ftools/fv
[49] http://heasarc.gsfc.nasa.gov/docs/xanadu/ximage/ximage.html

useful is the built-in support for calibration information from many missions when doing source detection and characterization.

Sherpa[50] is a high-level fitting and modeling application available in CIAO. Complex models can be defined for images. These models can include PSF effects and important detector properties contained in exposure maps.

6.3.4 Spectral-analysis software

Spectral analysis is a specialized task in X-ray astronomy requiring theoretical models appropriate to the waveband as well as computational and statistical techniques to deal with the instrumental response and small number of photons. Several large programs are available, each with their own advantages and disadvantages.

The most widely used program is XSPEC[51] from the HEASARC. This comes with a library of over one hundred theoretical models, many contributed by leading researchers as a way of disseminating their work. It is easy to include additional models and a web page is maintained of those made available to the community by their authors, but not yet installed in XSPEC as standard models.

The multi-dimensional fitting program Sherpa introduced in the previous section can also be used for spectral analysis. It contains its own models as well as importing those from XSPEC. It has powerful support for user-defined models and is well integrated with other CIAO tools. Sherpa uses the Python scripting language as its main interface and can be used in Python independently of CIAO, which allows flexibility when writing analysis scripts or parallelizing data processing.

The ISIS program[52] from the MIT Chandra HETG group was originally developed for high-resolution spectroscopy. It also includes the XSPEC model library. ISIS uses S-lang as a flexible scripting interface and has good support for parallel processing.

The SPEX program[53] from SRON in Utrecht specializes in collisional plasma models and high-resolution spectroscopy.

Two packages designed to aid in identifying lines using atomic physics data are Profit[54] and PINTofALE.[55] The former is a program with a GUI and the latter an IDL package.

---

[50] http://cxc.harvard.edu/sherpa   [51] http://xspec.gsfc.nasa.gov
[52] http://space.mit.edu/cxc/isis   [53] http://www.sron.nl/divisions/hea/spex
[54] http://heasarc.nasa.gov/docs/software/profit   [55] http://hea-www.harvard.edu/PINTofALE

6.3.5 Timing-analysis software

There is a wide variety of astronomical software available for the analysis of lightcurves, most of which can be used with X-ray astronomical data. A good source is the time-series analysis section of the StatCodes list.[56]

Two packages specifically written for X-ray astronomy are Xronos and SITAR. Xronos,[57] included in HEAsoft, is a collection of individual tools performing basic analysis such as plotting lightcurves, hardness ratios, colour–colour plots, calculating power spectra, searching for periods, and plotting lightcurves folded on a period. SITAR[58] is a set of functions and subroutines which can run in ISIS and offers many of the same options as Xronos. In addition, SITAR implements the Bayesian Blocks (Scargle, 1998) decomposition of a lightcurve.

6.4 Calibration data

Calibration data tend to change with time both because an instrument changes and also because the understanding of the instrument improves. Consequently, it is now standard to separate the data from the software so new calibrations can be included without updating the tools that use them. The calibration data files are stored in a database which ensures that the correct versions are used. There are two different database systems: that used for XMM–Newton and that used for everything else.

A good general source for all things related to calibration is the website of the International Astronomical Consortium for High Energy Calibration (IACHEC).[59]

6.4.1 CALDB

The HEASARC's calibration database scheme CALDB[60] has been further developed by the CXC. It consists of a file structure with index files and a directory tree for each mission. The file structure can be set up locally and filled with the files for the missions of interest. Alternatively, the CALDB can be accessed remotely from either the HEASARC or CXC sites. Remote access is easier for a few instances but some calibration files are large, particularly for Chandra, and it is better to set up the CALDB locally.

---

[56] http://www.astro.psu.edu/statcodes
[57] http://heasarc.gsfc.nasa.gov/docs/xanadu/xronos/xronos.html
[58] http://space.mit.edu/CXC/analysis/SITAR   [59] http://web.mit.edu/iachec
[60] http://heasarc.gsfc.nasa.gov/docs/heasarc/caldb

6.4.2 XMM–Newton

The Current Calibration File (CCF[61]) is the collection of all the XMM–Newton calibration files ever made public and is continuously updated. To ensure the most up-to-date calibration, the XMM–Newton team recommends that users mirror the CCF on their own systems. Each observation then needs its own index file to identify which calibration elements are required. This index file is created by the SAS tool `cifbuild`, as illustrated in Section 5 of the XMM–Newton ABC guide.

---

[61] http://xmm.vilspa.esa.es/external/xmm_sw_cal/calib

7

Statistics

ANETA SIEMIGINOWSKA

7.1 Introduction

Why do X-ray astronomers need statistics? Wall and Jenkins (2003) give a good description of scientific analysis and answer this question. Statistics are used to make decisions in science, evaluate observations, models, formulate questions and proceed forward with investigations. Statistics are needed at every step of scientific analysis. A statistic is a quantity that summarizes the data (mean, averages etc.) and astronomers cannot avoid statistics.

Here is a question asked by an X-ray astronomy school student:

> I wanted to know how many counts would be needed to get a good fit for a CIE plasma model with every parameter (save redshift) free. I was once told that it took 500–1000 counts to get a decent fit, but I couldn't remember if this assumed that metallicity is fixed. Can someone get a good fit for metallicity with so few counts?

What does "a good fit" or "a decent fit" mean, and what constitute "low counts" data? These expressions carry a definite meaning, but taken out of context are not precise enough. Is the question whether the total number of counts in the spectrum is "low" or whether the number of counts per resolution element is "low"? For example, a total number of counts such as 2000 could be viewed as "high," but it might be considered "low" if the total number of counts is divided by the number of resolution elements (for example there are 1024 independent detector channels in a Chandra ACIS spectrum). Also "high" is relative to the scientific question posed and the type of model. For example, a simple power-law model of a continuum could be well constrained by a Chandra spectrum with 500–1000 counts, but a plasma model with variable abundances would not.

7.2 The statistical underpinning of X-ray data analysis

The main goal of X-ray data analysis is to learn more about the physical properties of a source. Scientific questions are often formulated ahead of acquiring the X-ray data, i.e. during the time of target selection and application for the observations. So, data analysis is not performed blind, usually something is known about the observed source. Of course, discoveries can happen, even in the twenty-first century, and new data can bring unexpected results. One example is the discovery of large-scale X-ray jets in many Chandra observations of quasars.

When the new data arrive, statistics are employed to quantify the measurements. Typically, a parameterized approach to data modeling is taken wherein physical models with parameters (e.g. temperature and density of a hot medium emitting X-rays) are applied to the data. Finding and evaluating a model that best describes the observed data is the focus of data analysis. In general, this analysis can be divided up into different classes of statistical problem: (1) parameter estimation – defining a model and finding parameters that best describe the data; (2) confidence limits – how well the data constrain these parameters; (3) hypothesis testing – how well the model describes the data (goodness-of-fit) and whether it is the best model (model selection).

The choice of models is usually guided by prior knowledge and expectations of the observed source. Stars, for example, often have coronae filled with X-ray-emitting hot gas, so a thermal spectrum at a given temperature might be appropriate. If the source were a high-redshift quasar with a featureless continuum then a power-law model most likely describes the spectral slope. Statistics are used to find the model and associated parameter values that best represent the data. For each different set of parameter values, a likelihood function (the probability of a model) relates the model prediction to the observation. The model parameters that maximize the likelihood function are called the best model parameters. Usually the observed data are drawn from a Poisson distribution or, less frequently, a Gaussian distribution. Section 7.4 shows how to derive the likelihood function in these cases.

The model expression specifies a parameter space whose dimension is the number of model parameters. Two models can have disjoint parameter spaces, e.g. a power law with a slope parameter vs. a plasma model with parameters for temperatures and densities. Nested models are particularly important. In this case, the simpler model is a subset of a more complex model, e.g. a power law is a subset of a broken power law. The best model parameters define just one point in the model parameter space and finding it is non-trivial. Many optimization methods have been developed to find the best model parameters, although for models with many parameters there can be many possible solutions (so-called modes) that may need to be considered (see Section 5.2.4).

7.3 Probability distributions

Probability is usually interpreted in one of two ways. The Frequentist version is that probability measures the fraction of events of a particular type happening in a large number of identical trials. A paradigmatic example is repeatedly tossing a coin. In the Bayesian version probability is taken as a numerical measure of belief. This interpretation is adopted by astronomers when making statements of the likely range of quantities such as the Hubble constant.

The mathematical theory of probability, which is common to both interpretations, defines the probability distribution as follows. If x is a continuous random variable then $f(x)$ is its probability density function or, simply, probability distribution when:

(1) $\text{Prob}(a < x < b) = \int_a^b f(x)\mathrm{d}x$,
(2) $\int_{-\infty}^{\infty} f(x)\mathrm{d}x = 1$, and
(3) $f(x)$ is a single non-negative number for all real x.

The two probability distributions most widely used in X-ray astronomy are the Poisson and Gaussian (or normal) distributions.

- Poisson:

$$\mathcal{P}(n; \mu) = \frac{e^{-\mu}\mu^n}{n!} \tag{7.1}$$

is the probability of n events given a mean of μ. If μ is the "count rate," which is the average number of photons received from a source per unit time, then $\mathcal{P}(n; t\mu)$ describes the probability of receiving n photons in a given exposure time, t.

- Gaussian or normal:

$$\mathcal{N}(n; \mu, \sigma) = \frac{1}{\sigma\sqrt{2\pi}}\exp\left[\frac{-(n-\mu)^2}{2\sigma^2}\right] \tag{7.2}$$

is the probability of n events given a mean of μ and a variance of σ^2. If μ is the count rate then $\mathcal{N}(n; t\mu, t\sigma)$ describes the probability of receiving n photons in an exposure time t.

7.4 Parameter estimation and maximum likelihood

Suppose that $\{X_i, i = 1, N\}$ are X-ray data, independent and drawn from some probability distribution. For example, the X_i may be counts in bins in a spectrum or time series, or in pixels in an image. A source model described by the parameters Θ can be used to calculate a predicted M_i for each data point. The likelihood function is defined as:

$$
\begin{aligned}
\mathcal{L}(\{X_i\}) &= \mathcal{L}(X_1, X_2, \ldots X_N) \\
&= \mathrm{Prob}(X_1, X_2, \ldots X_N | \Theta) \\
&= \mathrm{Prob}(X_1 | M_1(\Theta)) \, \mathrm{Prob}(X_2 | M_2(\Theta)) \ldots \mathrm{Prob}(X_N | M_N(\Theta)) \\
&= \prod_{}^{N} \mathrm{Prob}(X_i | M_i(\Theta))
\end{aligned}
\tag{7.3}
$$

where $\mathrm{Prob}(X_i | M_i(\Theta))$ is the probability that the ith data point has value X_i given a model-predicted value of M_i calculated for parameter values Θ. Finding the maximum likelihood means finding the parameters Θ_0 that maximize $\mathcal{L}$.

7.4.1 Poisson data

Suppose now that the X_i follow a Poisson distribution. The expression for the likelihood becomes:

$$
\mathcal{L}(\{X_i\}) = \prod_{}^{N} \mathcal{P}(X_i; M_i(\Theta))
\tag{7.4}
$$

As an example of calculating the likelihood suppose that the X_i come from an X-ray spectrum. The M_i (expected model counts in detector channels) are evaluated according to Equation 4.10 using the source spectral model, the detector redistribution matrix (RMF) and telescope effective area (ARF). For a power-law function $m(E) = A E^{-\gamma}$, where A is the normalization in photons/cm^2/sec/keV, E is photon energy in keV and γ is the photon index, the predicted number of counts, M_i, in a detector channel i is then calculated as:

$$
M_i = T \int \mathrm{RMF}(i, E) \cdot \mathrm{ARF}(E) \cdot A E^{-\gamma} \cdot dE
\tag{7.5}
$$

This integral is solved in the XSPEC or Sherpa spectral fitting programs to calculate model-predicted counts M_i for given values of the parameters (A, γ). Assuming standard Chandra ACIS calibration and power-law parameters of $\gamma = 2$ and $A = 0.001$ photons/cm^2/sec/keV then the model-predicted counts in the ACIS channels $i = (10, 100, 200)$ are $M_i = (10.7, 508.9, 75.5)$. Suppose the observed counts in these channels are $X_i = (15, 520, 74)$. The Poisson likelihood is then:

$$\mathcal{L}(\{X_i\}) = \prod_{i}^{N} \mathcal{P}(X_i; M_i(A, \gamma))$$ (7.6)
$$= \mathcal{P}(15; 10.7)\mathcal{P}(520; 508.9)\mathcal{P}(74; 75.5)$$
$$= 3.37 \times 10^{-5}$$

where individual Poisson probabilities are calculated given the observed data in these three channels.

In a typical application the model is evaluated in all the detector channels. Finding the maximum likelihood means finding the best set of parameters (A, γ) that maximize the Poisson likelihood. The likelihood is calculated for many sets of the parameters (A_j, γ_j) and the set that gives the maximum value of $\mathcal{L}$ is selected as the most likely description of the observed source. Many numerical methods have been developed to make this iterative process efficient. van Dyk *et al.* (2001) describe Monte Carlo algorithms specific to X-ray spectra.

In many applications a log-likelihood approach is used. Taking the natural log of the Poisson likelihood and using simple algebra gives:

$$\ln \mathcal{L} = \sum_{i}(X_i \ln M_i - M_i - \ln X_i!)$$ (7.7)

Cash (1979) defined the C statistic using this log-likelihood function and multiplying it by -2.

$$C = -2 \ln \mathcal{L} = \sum_{i}(X_i \ln M_i - M_i - \ln X_i!)$$ (7.8)

The term $\ln X_i!$ is independent of the model so can be ignored in the optimization leaving:

$$C = 2 \sum_{i}^{N}(M_i - X_i \ln M_i)$$ (7.9)

7.4.2 Gaussian data

A few, usually non-imaging, detectors generate data whose variation is not Poisson but Gaussian with a sigma that is known and independent of the data. In this case the maximum likelihood estimator is:

$$\mathcal{L}(\{X_i\}) = \prod_{i}^{N} \mathcal{N}(X_i; M_i, \sigma_i)$$ (7.10)

so:

$$-2\ln\mathcal{L} = \sum_i^N \mathcal{N}(X_i; M_i, \sigma_i)$$

$$= \sum_i^N \frac{(X_i - M_i)^2}{\sigma_i^2} - 2\sum_i^N \ln\frac{1}{\sigma\sqrt{2\pi}} \qquad (7.11)$$

Since the second term is independent of the model this reduces to the widely used statistic:

$$S^2 = \sum_i^N \frac{(X_i - M_i)^2}{\sigma_i^2} \qquad (7.12)$$

which is distributed as χ^2. The statistic itself is often referred to as χ^2 but we prefer to reserve that name for the probability distribution itself.

The S^2 statistic is often used in cases where it is not the correct likelihood estimator, i.e. the variation in the data is not Gaussian or σ is not known. Usually this is because S^2 is particularly easy to modify to include background, provides a convenient goodness-of-fit test (see Section 7.6) and X-ray data are usually binned. Note that using S^2 with Poisson data (where σ_i^2 is estimated by X_i or M_i) can lead to biased results (Section 7.7.1).

7.4.3 Likelihood and Bayesian posterior probability

In the Bayesian framework, the probability of a model given the data (the posterior probability) is described by a combination of the likelihood and priors:

$$\text{Prob}(\Theta|\{X_i\}, I) = \text{Prob}(\{X_i\}|\Theta, I)\text{Prob}(\Theta|I) \qquad (7.13)$$

Formally, this equation also contains a normalizing factor which ensures that the probability is between 0 and 1. We ignore this factor in the following description. The left-hand side of Equation 7.13 is the posterior probability, i.e. the probability of a model with parameters Θ given the observed X-ray counts, $\{X_i\}$. The first term on the right-hand side is the likelihood function, which is the probability of the data, $\{X_i\}$, originating from a model with parameters Θ. In the case of Poisson-distributed data this term is Equation 7.4. The second term on the right-hand side of the equation is the prior, which is the probability of the model with parameters Θ given previous knowledge, I, about the source. Examples of such previous knowledge are the type of emission model, or a range of parameters that are applicable to the source and probability distributions of these parameters. The prior can have its own parameters. For instance, the prior may be that each model parameter, θ_k, is described by a Gaussian probability distribution with its own μ_k and σ_k. In this case, Equation 7.13 can be rewritten, taking natural logarithms:

$$\ln \text{Prob}(\Theta|\{X_i\}, I) = \ln \mathcal{L}(\{X_i\}|\Theta, I) + \sum_k \ln \mathcal{N}(\theta_k; \mu_k, \sigma_k) \quad (7.14)$$

The posterior distribution can be determined very efficiently using MCMC (see Section 5.2.4.2) methods. Each step in the MCMC chain is calculated by drawing model parameter values randomly from their prior distributions, calculating the likelihood given this set of model parameters, then evaluating the posterior probability using Equation 7.13. If the Metropolis–Hastings criterion is being used then the step will be accepted or rejected based on the likelihood ratio to the last step.

7.5 Confidence bounds

Maximum-likelihood estimators such as C or S^2 are used to find the best model parameters given the X-ray data. However, just knowing the best parameters is not generally scientifically useful. Knowing how well the parameters are determined is also important. This depends on the quality of the data: signal-to-noise ratio, total number of counts, etc.

The MCMC methods provide a complete multi-dimensional posterior probability distribution for all the model parameters given the data. These are not easily published or even, for large numbers of parameters, visualized. However, posterior probability distributions do provide the full information about the parameters and they can be displayed using scatter or contour plots. The most common way to report a precision to which a parameter is determined is to use a confidence (or credible) bound or region. An example is to state that there is a 90% probability that the true power-law index in an AGN spectrum lies between 1.6 and 1.8. This is calculated from the complete posterior probability distribution by integrating over all other parameters then choosing a range that encompasses 90% of the total probability. The second step is not unique since there is an infinite number of ways of choosing such a range. The two most common of these ways are to choose either the smallest range that covers 90% of the probability or the range that is centered on the best parameter value.

Often, confidence ranges are required on derived quantities such as fluxes and equivalent widths. These can be estimated by randomly drawing sets of parameter values from the posterior probability distribution and calculating the derived quantity for each set.

This complete analysis was not possible before the advent of modern computers so analytic approximations were developed and are still often used. Avni (1976) showed that for the S^2 statistic the confidence region can be estimated by searching for parameter values such that

$$S^2(\alpha) = S^2_{\min} + \Delta(\nu, \alpha) \tag{7.15}$$

where α is the significance level and ν the number of degrees of freedom (i.e. the number of channels minus the number of parameters). For example, $\Delta = 2.71$ for the 90% confidence region for a single parameter. Cash (1979) showed that the same approximation holds for the C statistic. Finding the parameter value that gives the required value of S^2 or C involves searching around the best fit and may be still be numerically intensive, particularly if parameter space is complex.

A further simplification, which is sometimes used, is to calculate the matrix of second derivatives of the statistic with respect to the model parameters and then use the diagonal elements to estimate confidence regions (they will always be symmetric). In this case the correlations between parameters are not considered and the confidence bounds may not be correct. This is especially problematic for highly correlated parameters, such as the photon index and absorbing column in the popular absorbed power-law model applied to X-ray spectra.

7.6 Hypothesis testing and model selection

Data analysis does not end with finding the best-fit model parameters. If the fit is not good because the model is incorrect then the parameter values mean nothing. On the other hand, the fit may be good but there may be other models which work as well or better. These issues are resolved using hypothesis testing.

The classical hypothesis test determines whether some hypothesis, H_0 (called the null model and usually formulated to be rejected), is true given the observations. An example of H_0 is: "there is no point source at position X, Y." There are two possible types of error that can be made when doing such a test: either the null model is rejected when it is really true (false positive or type I error) or the null model is accepted when it is really false (false negative or type II error). It is impossible to choose a test that simultaneously minimizes both these errors. The standard statistical approach is to limit the probability of type I error to below some significance level α (5%, say). The power of the test is then defined as the probability of (correctly) rejecting H_0 when it is false. The power can be written as $1 - \beta$, where β is the probability of a type II error of accepting a false H_0.

The first step in performing an hypothesis test is to choose a test statistic, T, with as much power as possible. The next step is to specify, a priori, the significance level, α. The test statistic, T, is then calculated for the observed data. For a few special cases of test statistic, which have a probability distribution independent of the model being tested, it is then possible immediately to calculate the p-value, which is defined as the probability that the value of T be as extreme as observed. More generally, this probability is evaluated by performing simulations. Many data sets are simulated assuming the null model and a value of T calculated for each one. This set of T values constitutes an empirical estimate of the probability distribution with the p-value being the fraction of $T_{\text{simulated}}$ which are greater than or equal to T_{observed}. Finally, H_0 is rejected if the p-value is less than α.

7.6.1 Goodness-of-fit

In a classic paper, Pearson (1900) showed that

$$S^2_{\text{Pearson}} = \sum_{i}^{N} \frac{(X_i - M_i)^2}{M_i} \qquad (7.16)$$

is approximately distributed as χ^2 with $N - 1$ degrees of freedom. This formula defined the χ^2 goodness-of-fit test. Subsequent experience established that individual X_i should not be too small and a well-known rule of thumb was adopted that they all be ≥ 5. It is straightforward to determine S^2_{Pearson} then calculate the p-value using the χ^2 distribution. An even simpler test is that S^2_{Pearson} should be approximately equal to the number of degrees of freedom. Note that this test statistic is not the same as the S^2 defined for Gaussian-distributed data in Section 7.4.

While Pearson's goodness-of-fit statistic is the one most widely used in X-ray astronomy there are many others available. Examples include Kolmogorov–Smirnov, von Mises and Anderson–Darling (see e.g. Babu and Feigelson 1996). Note that most of these tests are non-parametric and typically used to compare samples. They are rarely used to assess goodness-of-fit. It is worth noting that these tests were developed before modern computing so that they are designed to have probability distributions that are approximated by known, tabulated distributions. With more computing power available it is possible to do tests entirely using simulations.

7.6.2 Model selection: likelihood ratio test

The likelihood ratio (LR) test and F-test have been widely used in X-ray analysis (see for example Cash 1979, Protassov *et al.*, 2002). Both tests compare a simpler model (e.g. a power law) with a more complex model (e.g. a broken power law). For these tests to be valid the following conditions have to be true:

- the simpler (null) model is nested within the other more complex (alternative) model, i.e. it is a subset of the complex model;
- the extra parameters of the alternative model have Gaussian (normal) distributions that are not truncated by parameter space boundaries.

These tests can be used to compare a broken power-law model to a power law, but not to compare a power-law model and a thermal model because, in this case, one model cannot be nested within the other one. The second condition has often been violated in testing for an emission line in X-ray spectra. In this case, the null model does not have an emission line while the alternative model does. One of the parameters of the alternative model is thus the emission-line flux, which is truncated by the parameter space boundary because the flux must be non-negative and the null model corresponds to the flux being zero. This is not merely a technical requirement, but it can lead to significant errors in scientific inference. Protassov *et al.* (2002) give detailed descriptions of such incorrect applications of these tests and provide alternative tests that should be used.

Simulations and calculations of p-values[1] are often the best approach to compare models. As an example, consider calculating *p*-values to test for the presence of an emission line. This is illustrated by Figure 7.1, which was generated using the following steps (see also Figure 7.2): (1) fit the observed data with a simple (null) model; (2) fit the observed data with an alternative model and record the LR between the alternative and null model fits; (3) generate a large number (~500) of simulated data sets assuming the null model; (4) fit each simulated data set with the null and alternative model; (5) calculate the LR for each simulated data set; (6) make a histogram of the LRs and plot as shown in Figure 7.1; (7) compute the p-value, i.e. the probability of the observed LR given the distribution of LR in the simulations; (8) decide whether the null model is rejected or accepted. Usually a p-value < 0.05 would reject the null and accept the alternative model. In the simulations shown in Figure 7.1 the null model would be selected and the evidence for a complex model rejected.

---

[1] Posterior predictive p-values are Bayesian analogs.

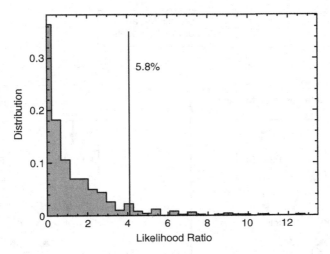

Fig. 7.1 The null distribution of LR test statistic. The histogram shows the LR values calculated from simulating the null model (a power law) and fitting the simulated data with a null and an alternative model. The vertical line indicates the observed value of LR and 5.8% is a p-value for these data, i.e. a fraction of LR values that are located to the right of the line

7.7 Statistical issues

7.7.1 Bias

Statistical bias occurs if the expected value of an estimator for a parameter value differs from the true value. Bias can affect X-ray analysis and is best tested for using simulations. As an example, consider using statistics that assume an underlying Gaussian distribution on data which are Poisson. The bias is easily shown using simple simulations performed in Sherpa or XSPEC.

This is illustrated in Figure 7.3, which was made as follows. One thousand simulated Chandra X-ray spectra were generated based on an absorbed power-law model with three parameters (an absorption column, a photon index, and a normalization) and the standard instrument calibration files (RMF/ARF). These simulated X-ray spectra thus contained the model predicted counts with Poisson noise. Each spectrum was then fitted three times, using different fit statistics, to the absorbed power-law model and the best-fit parameter values determined. One time used S^2, another S^2_{Pearson}, and the third C.

The figure shows the distribution of the photon index parameter obtained from the fit to these high signal-to-noise spectra which were generated with a photon index of 1.28. The S^2 bias is evident in this analysis. While

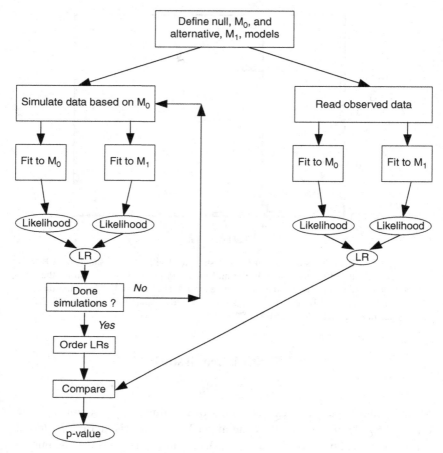

Fig. 7.2   A flow diagram showing how to use the LR test decide whether to reject the null model in favor of an alternate. The final comparison step is illustrated in Figure 7.1

C statistics based on the Poisson likelihood behave well, S^2_{Pearson} underestimates and S^2 overestimates the photon index.

A number of other variants of S^2 have been suggested using different denominators (e.g. the Gehrels, Churazov, Primini weighting options in XSPEC or Sherpa). If any of these are used they should always be checked for significant bias by using simulations.

7.7.2 Source detection and upper limits

Source detection presents some interesting statistical issues. The decision about whether or not a source has been detected requires a statistical measure of the

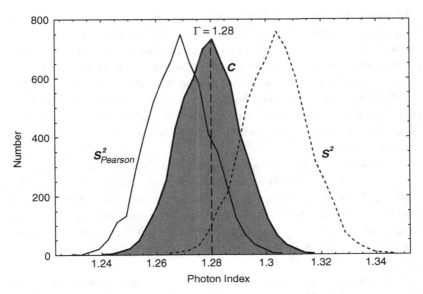

Fig. 7.3   Distributions of a photon index parameter γ obtained by fitting simulated X-ray spectra with 60 000 counts and using the three different statistics: S^2_{Pearson}, S^2 and C (i.e. the Poisson likelihood) statistics. The true value of the simulated photon index is marked with a dashed line and it was set at $\gamma = 1.28$

source significance and a detection threshold. If the result is a non-detection then an upper limit may be required. Observations can be affected by high or low background counts, background flares, or instrumental features (e.g. hot pixels, node, or chip boundaries) and all these effects need to be taken into account. Depending on the PSF size there could also be many unresolved sources contributing to the detection region. These effects are typically included in the standard source-detection algorithms (see Section 5.4.1).

A bright point source is relatively easy to detect and the standard characterization (e.g. signal-to-noise ratio) can be used. However, detections of faint sources (or upper limits) involve statistical tests to establish the source significance. In this case simulations are used to calculate the probability of a background fluctuation that would generate a detected signal. We illustrate this with an example from Chandra, which has a very low background.

To obtain the significance of detecting five counts from a source in a 5 ksec Chandra observation do the following simulations. After selecting a background region free of obvious sources, measure the number of counts, N_b, in the region. Assume, for simplicity, that the background is uniform across the source and background regions. The observed counts are generated in a Poisson process with mean given by N_b, so randomly draw a number of background counts assuming this distribution. Rescale the drawn counts by the ratio of the source and background areas. This gives a predicted number of background counts in the source region. For each of these predictions calculate the p-value, i.e. the significance of obtaining the observed five counts given the predicted background. The final significance is a mean of all the simulated p-values. It was 6.1×10^{-7} in one of our Chandra observations (with five counts in the source region and six counts in the background region, which was 50 times larger than the source region) indicating that the counts originated in a source, not in the background.

If the test is for a known source position then the calculation is as above. However, if the entire field is being searched for sources then the probability needs to take into account the number of tests being performed. In this case, the entire field should be simulated many times and the mean number of false positives found. It is then only possible to state that some number of sources have been detected of which a fraction will be false positives.

Statistical issues associated with calculations of upper limits are discussed in a recent paper by Kashyap *et al.* (2010). The upper limit is closely related to a detection process. It is usually possible to define a detection threshold in the observation based on the background intensity and exposure time. An upper limit is then defined as a minimum (or maximum) intensity that a source could have had to be detected (or non-detected) with a required probability given by the detection algorithm. Note that this is different from the confidence regions described in Section 7.5.

7.7.3 Background subtraction

X-ray data are collected as individual events. Both source and background counts are contained within the assumed source region, which is usually matched to the PSF, so the background contribution needs to be accounted for in any analysis. In the high counts case with a low background contribution, the background is typically subtracted and the error propagated according to standard theory:

$$N_{\text{net}} = N_{\text{S}} - \frac{A_{\text{S}} t_{\text{S}}}{A_{\text{B}} t_{\text{B}}} N_{\text{B}} \tag{7.17}$$

where N_{net} is the net source counts, N_{S} the observed counts in the source region, N_{B} the observed counts in the background region, A_{S}, A_{B} are source and background regions respectively and t_{S}, t_{B} are source and background exposure times.

In the low-counts regime, the treatment of the background is tricky since subtracting the background can lead to negative(!) counts. This is where Poisson likelihood and therefore C should be used. However, while the difference of two Gaussian variables is another Gaussian variable, the difference of two Poisson variables does not have a simple distribution. In this case, the background cannot be subtracted but the source and background regions analyzed simultaneously. Examples of this are given in Section 5.2.4.4 and Chapter 8.

7.7.4 Rebinning

Rebinning or grouping of low-counts data is sometimes used in X-ray data analysis to increase the number of counts per bin. This is not a good idea in general as it leads to the loss of information. For example, in the analysis of spectra, an emission line or absorption edge could disappear during rebinning. However, if there are only a handful of counts and as long as no sharp features are expected then rebinning provides a quick and easy way to perform preliminary analysis. Final analysis, however, should use the C statistic so rebinning is not required. Rebinning options for plotting can only be used to create attractive plots while not losing any usable data.

7.7.5 Systematic errors, calibration, and model uncertainties

Observations usually have systematic errors in addition to statistical errors. The most common way to include these errors is to add the statistical and systematic errors in quadrature. However, such an approach does not account for non-linear errors, such as those present in most calibration files. Drake *et al.* (2006) show how these errors affect the confidence of the best-fit parameters and that systematic calibration errors dominate the error budget in very high signal-to-noise-ratio observations. In this case, the constraint on the best-fit parameters is limited by the size of the systematic errors and simulations should be used to estimate the parameter confidence regions (Drake *et al.*, 2006).

Another class of uncertainties is related to physical models. For example, in the plasma emission models for spectra, the energies and strengths of line transitions are not precisely known. Such uncertainties should be included in any analysis; however, at present they are not propagated and standard fitting packages do not take them into account.

8

Extended emission

KIP D. KUNTZ

8.1 Introduction

This chapter describes X-ray data reduction for extended sources; the process required to isolate an image of the source from the background or to isolate the source spectrum from all of its backgrounds. We begin by describing the ways in which data are less than ideal and what general methods are employed to work around these problems. We then walk through a typical reduction, posing and discussing the questions that need to be considered with each step.

Study of diffuse sources is usually done with imaging spectrometers such as position-sensitive proportional counters and CCDs. Combining spectral and imaging (photometric) data increases the versatility of the analysis, but that versatility comes at the price of more complex procedures, which require simultaneous analysis of both the spectra and the photometry.

A brief consideration of photometry demonstrates why this is the case for non-ideal detectors. First, a region is defined that includes the source and a second "background" region that is free of sources. For an ideal detector, the surface brightness in the background region (in counts/pixel) can simply be determined and subtracted from the source region. Several modifications are required for non-ideal detectors where the response and the point spread function (PSF) may vary with position.

(1) If the source region is not large compared to the PSF, then the flux in the source region must be corrected for the amount of source flux falling outside the region. Such a correction is not often required for extended emission, but it can be important (and quite difficult) for large PSF instruments such as the Suzaku XIS.

(2) If the source region is not small compared to the instrument's FOV (and sometimes even when it is) the instrumental response may vary significantly over the source region. Complicating the issue, the variation in response over the

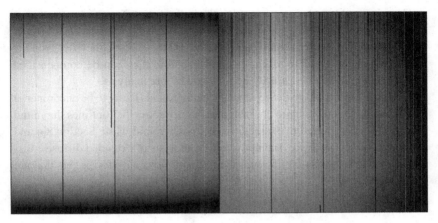

Fig. 8.1 [Left] The Chandra effective area at 1.0 keV as a function of position on the S3 chip. [Right] The Chandra effective area at 6.0 keV as a function of position on the S3 chip. Both images have been stretched from 80% to 100% of the maximum response

detector is usually a strong function of energy (see Figure 8.1). This variation is not a problem when dealing with a relatively narrow bandpass. However, the low count rate and relatively poor spectral resolution typical of current X-ray telescopes/detectors encourage studies over broad bandpasses (> 0.5 keV wide) for which the differential variation with energy may be significant. This complication is usually handled by calculating an emission-weighted mean response over the source region. This solution is not always adequate; consider a large, radially symmetric source (e.g. a cluster) centered on a detector with a (typically) radially symmetric variation in the response.

(3) Given that the response may vary over the source region, the variation between the source and background regions will be even greater. This complication is compounded by the following: (a) X-ray background rates are low (but not negligible) so statistically significant backgrounds are, by necessity, extracted over large regions of the detector. (b) The source is usually placed at the center of the detector. If it is at all large, the background regions must be placed near the edge of the detector/FOV where the response tends to differ from that of the center. (c) The background may have an entirely different spectral shape than the source. For these reasons it is not, in general, a good idea to subtract the background directly from the source. This is particularly true as some background components show spatial variation that is not correlated with the variation in the instrument response; for example, the particles are generally not vignetted in the same way as the photons.

Thus, it should not be surprising that the bulk of the effort (and uncertainty) in the analysis of diffuse/extended emission is in the characterization and "removal" of the backgrounds (note the "s"!). In general, the backgrounds must be modeled to some extent or another before they can be compensated for, and they may not necessarily be removed before analysis.

Section 8.2 describes in detail all of the backgrounds likely to be encountered and the methods by which to characterize them. Section 8.3 outlines the initial reduction steps required to characterize the background while Section 8.4 and Section 8.5 discuss the spectral and imaging analysis of extended diffuse emission. Finally, Section 8.6 considers the issue of mosaicing for targets that are larger than the FOV.

8.2 Backgrounds and foregrounds

Consider the observation of an interesting X-ray-emitting object. In addition to X-ray photons from the object, the detector records X-ray photons from a number of other sources that typically fill the entire FOV, such as the diffuse X-ray background, Galactic emission, and heliospheric emission. These are the *cosmic* backgrounds. In orbit, the detector is bathed in energetic particles. These particles can either interact with the detector directly, sometimes producing a signal indistinguishable from that produced by X-rays, or interact with the materials around the detector, producing fluorescent and hard-target bremsstrahlung X-rays that may then strike the detector. These are the *instrumental* backgrounds.

This discussion will address the background problem under the assumption that the source of interest fills the entire FOV and that the backgrounds have to be constructed "from first principles." Most analyses do not require such extreme measures, but many of these techniques are needed when the background must be measured far from the source of interest or the background is highly variable. We will concentrate on spectral analysis under the assumption that photometry is just spectral analysis with very large binning.

8.2.1 Instrumental backgrounds

The particle background is the term used to describe the background recorded by the instrument when it is not exposed to cosmic X-rays. It is due to charged particles interacting with the detector, or interacting with material around the detector and producing X-rays that then strike the detector. Chandra measures the particle background for ACIS by taking exposures when the detector is under

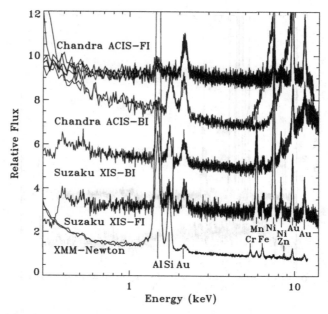

Fig. 8.2 Typical particle background spectra from current missions. Prominent lines are identified. Curves have been normalized and offset for clarity. *FI:* Front-side illuminated; *BI:* back-side illuminated

a shield in the "stowed position." The XMM–Newton satellite, like ROSAT, closes the filter wheel for such measurements; ASCA and Suzaku observe the night side of the Earth under the assumption that any X-ray emission between the spacecraft and the Earth will be insignificant. Figure 8.2 shows particle background spectra for a number of current missions.

The spectrum of the particle background consists of a continuum (primarily from the direct interaction of the particles with the detector) and lines (due to X-ray fluorescence of the surrounding material). For XMM–Newton, the sensitivity of the detector to the continuum varies across the instrument, and different parts of the detector "see" different fluorescing materials. The response of the Suzaku XIS detectors and the Chandra ACIS detectors appears to be much more uniform. The shape of the particle background spectrum varies with time, as the particle environment varies with time and the location in the orbit. For most current missions, standard particle background spectra are available. These are easily scaled to the strength of the background during the source observation. The instrumental response to cosmic X-rays usually becomes insignificant at higher energies while the response to particles does not. Thus, the observed strength of the spectrum at energies above $\sim$10 keV

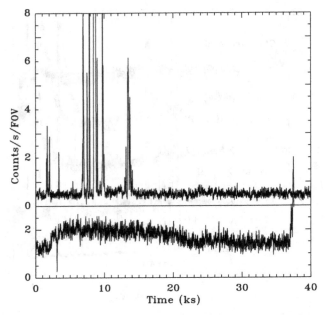

Fig. 8.3 Lightcurves showing soft-proton contamination in XMM–Newton observations. The top plot is a typical lightcurve; the soft-proton contributions are strong but of limited duration. The bottom plot shows that the soft-proton contribution can be relatively constant for an extended period of time; a flat lightcurve for a short (10 ks) observation is no guarantee of low soft-proton contamination

(for typical CCD detectors) can be used to normalize the standard spectrum. A somewhat more complicated method has been developed for XMM–Newton (Kuntz and Snowden, 2008).

The soft-proton contamination (SPC): One of the unpleasant surprises of the Chandra and XMM–Newton missions was that the mirrors focus low-energy protons ($\sim$150 keV) onto the detectors. These soft protons are modulated by the Earth's magnetic field, so the soft-proton flux depends upon time, spacecraft location, and pointing direction. Soft-proton events have strongly variable strengths on timescales from seconds to hours. Figure 8.3 shows XMM–Newton lightcurves of observations of "blank" sky. For most observations, soft-proton enhancements are readily distinguishable as strong positive features in the lightcurve. However, some soft-proton events show low amplitudes and small variation, so there is always a chance of residual SPC after the obvious events are removed from the lightcurve. Among the current missions, XMM–Newton is unique in having a method to measure this residue (de Luca and Molendi, 2004; Kuntz and Snowden, 2008). The spectral shape of the SPC for Chandra and XMM–Newton can be described as a rather flat power law (perhaps broken or cut off) which is usually clearly apparent at $E > 3$ keV. For most studies, the SPC can be fitted simultaneously with the source spectrum. However, because the soft protons are not photons, the standard instrument response should not be applied but a diagonal response matrix should be used instead.[1] Similarly, the soft-proton vignetting (i.e. their distribution over the detector) is very different from the photon vignetting,[2] so even studies of point sources must be done carefully if there is a substantial SPC. The ROSAT and ASCA missions did not experience the same types of flares, and Suzaku is similarly immune to this problem, implying that their low Earth orbits are relatively free of SPC; a study of SPC in XMM–Newton suggests that it is most prevalent between the bowshock and the magnetopause, far above low Earth orbits.

8.2.2 Cosmic backgrounds

The extragalactic background: In the 0.1–10.0 keV band, the extragalactic background is composed almost exclusively of unresolved AGN. Whether there is any emission remaining after removing all of the AGN remains an active area of research, particularly at lower energies. The spectral shape and the normalization of the extragalactic background depend upon the extent to which the AGN are removed, as the fainter AGN tend to have flatter spectra than the bright AGN. Although the normalization of the extragalactic background does seem to vary (Cappelluti *et al.*, 2007), for typical exposures the unresolved AGN emission between 0.1 and 7 keV can be characterized by a power law of photon index $\sim$1.4 and a normalization of 10 keV/cm^2/s/sr/keV (Chen *et al.*, 1997) before the removal of resolved sources.

The Galactic foreground: There are at least two components at high Galactic latitudes, and even more in the disk. The *Local Hot Bubble* (LHB) is an irregular region surrounding the Sun with a radius of 100–200 pc that is deficient in neutral gas and is thought to be filled with gas at a temperature of 10^6 K (Snowden *et al.*, 1998). The LHB emission is seen, unabsorbed, in every observation. The *Galactic halo* has temperatures $1–3 \times 10^6$ K and typically emits beyond the bulk of the neutral disk gas, and is therefore absorbed to a variable extent. The ROSAT All-Sky Survey (RASS) shows that the halo spectrum varies across the sky (Kuntz and Snowden, 2000) and Chandra, XMM–Newton, and Suzaku results confirm this variation.

---

[1] The photon response function usually has small-scale features that do not appear in the particle background spectrum. Thus, the source spectrum cannot be modified to suit the response.

[2] At least for XMM–Newton; the response for Chandra is not as well studied though a preliminary analysis suggests the same.

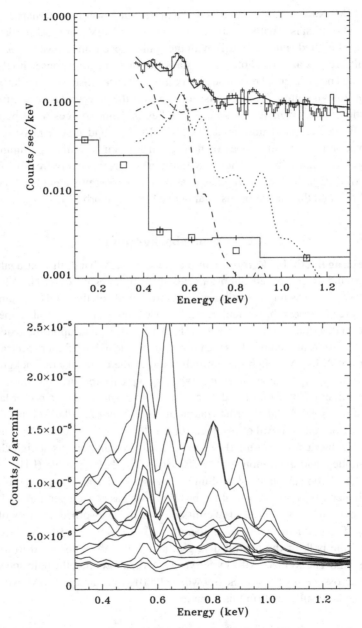

Fig. 8.4 [Top] Typical high Galactic latitude spectrum, taken with XMM–Newton, decomposed into LHB (dashed), halo (dotted), and background AGN (dot–dashed). The lower line (data) with boxes (model) is the RASS spectrum for the same region. [Bottom] The variety of Galactic foreground spectra

At CCD spectral resolution, the LHB and the Galactic halo can be well fitted by collisional ionization equilibrium models; the emission is predominantly in lines and the bulk of the emission falls below 1.5 keV. Figure 8.4 shows the decomposition of a typical Galactic-foreground spectrum. If the source spectrum has significant flux below 1.5 keV, then the Galactic foreground must be treated with caution; measurements taken as close as 5° away may not adequately represent the Galactic component in the FOV, nor will averages over large numbers of "blank sky" fields. The best solution is to fit the Galactic foreground simultaneously with the source spectrum using the form

$$\mathrm{APEC_{LHB}} + e^{-\tau}(\mathrm{APEC_{halo}} + \mathrm{APEC_{halo}}) \tag{8.1}$$

for the Galactic emission. The MEKAL models can be substituted for the APEC models. The temperatures of the components should be allowed to vary; typically $kT_{\mathrm{LHB}} \sim 0.1$ keV and $kT_{\mathrm{halo}} \sim (0.1, 0.2)$ keV, where the softer halo component may have a very low normalization. Since the bulk (but not all) of the emission from a $kT \sim 0.1$ KeV plasma is below the XMM–Newton (or Chandra) bandpass, these soft components must be constrained below 0.3 keV to obtain a realistic fit in the 0.3 to 10 keV bandpass. Such a constraint can be placed using a spectrum extracted from the RASS[3] from a $0°.5 \leq R < 1°$ annulus around the source position. This RASS spectrum is then fitted simultaneously with the source spectrum (Henley *et al.*, 2007; Kuntz and Snowden, 2008). Note that ROSAT pointed data are not useful for this purpose since they have not had the heliospheric emission removed at all, while the RASS has had the time-variable portion (discussed next) of the heliospheric emission removed.

Heliospheric and geocoronal emission: The solar wind contains ions which emit X-rays when they interact with neutral H and He and exchange an electron. The importance of this emission mechanism to X-ray astronomy had been unknown until ROSAT observations of comet Hyakutake revealed it to be very bright in X-rays (Lisse *et al.*, 1996). Similar emission, termed solar-wind charge exchange (SWCX), is produced by interaction of the solar wind with the neutral interstellar medium (ISM) that flows through the Solar System (the heliospheric emission) as well as interaction with neutrals in the Earth's exosphere (the geocoronal emission). The SWCX emission was observed by ROSAT (although the origin was unknown at the time) as background variations on hour to week scales called "long-term enhancements." ROSAT observations of the dark side of the Moon suggests that at least part of this emission was produced by solar-wind interaction with exospheric (i.e. between the Earth and

---

[3] The ROSAT data can be obtained easily from the X-ray background tool at http://heasarc.gsfc. nasa.gov/cgi-bin/Tools/xraybg/xraybg.pl

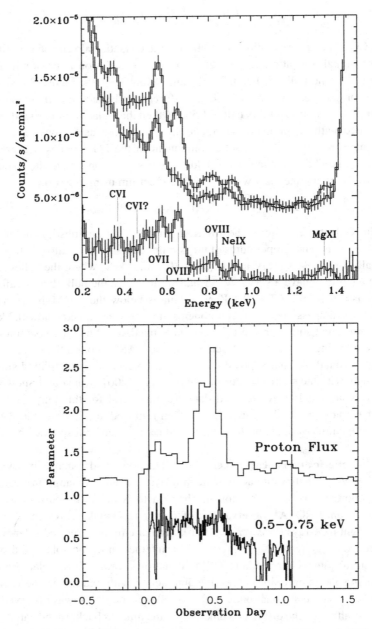

Fig. 8.5 [Top] Two different XMM–Newton observations of the same field (in this case the HST Deep Field), showing the variability of the SWCX contamination. The bottom line is the difference offset by -2.0×10^{-6}. [Bottom] The ACE proton data (top) and the 0.5–0.75 keV lightcurve for the observation interval more strongly contaminated by SWCX emission. The difference in the curves is due to geometric effects

the Moon) neutral atoms; Lallement (2004), Koutroumpa *et al.* (2009), and Robertson *et al.* (2009) have shown that the contribution from interaction with the neutral ISM within the Solar System produces a substantial part (but not all) of the emission previously attributed to the LHB.

The amount and spectral shape of heliospheric emission are still poorly understood. Charge-exchange emission is composed entirely of lines; however, because the ionization structure of the solar wind varies, the line ratios vary as well. Figure 8.5 shows two different observations of the same location, one of which was strongly affected by SWCX emission. The SWCX emission affects the important O VIII and O VII lines, and can sometimes be detected by the presence of Ne IX and Mg XI lines. Short SWCX enhancements (a few hours in duration) are usually well correlated with strong increases in the solar-wind proton flux, so examination of ACE proton data can indicate time periods likely to show SWCX contamination. Unfortunately, the converse is not true, as SWCX contamination can occur even though nothing unusual is seen in the ACE data.

Observations through the sub-solar point of the magnetosheath may also be strongly contaminated by SWCX emission. There is as yet no way of determining the *total* SWCX contribution.[4]

Exospheric emission: Spacecraft in a low Earth orbit, such as ROSAT or Suzaku, often observe quite close to the limb of the Earth. In such cases, aurorae and solar X-rays scattered from the Earth's atmosphere can also contaminate the data. There are usually mission-specific programs for characterizing and removing these backgrounds.

Your observation: Of these many backgrounds, which do you have to worry about? You will always have to remove the particle background. Chandra and XMM–Newton will always have some soft-proton contamination while ROSAT and Suzaku may have exospheric contamination. The importance of the extragalactic background, the Galactic foreground, and SWCX depends upon the location and spectral shape of your object. It is always useful to look at the RASS to see what Galactic emission might be in that direction. You can often use the RASS count rates and simple models of the Galactic foreground to estimate the extent of the problem for your object.

8.3 Initial analysis

Lightcurves: To determine the extent of the contamination by time-dependent backgrounds, it is best to create lightcurves for the emission from the entire

[4] The author is developing an SWCX "toolbox" which will help identify observations that are likely to be seriously contaminated. It will be available through the GSFC web pages.

FOV, excluding any bright variable sources. Any period in the lightcurve with enhanced emission should probably be removed; most diffuse emission has a surface brightness on the order of, or smaller than, the background so the contribution to the noise by periods of enhanced emission is often more detrimental than the loss of the source data from that period. Background enhancements may be due to increases in the particle background, often at the beginning or the end of the exposure when the spacecraft is exiting or entering the particle belts, though short (~15 minute) enhancements can occur at any time. For spacecraft in low Earth orbit (ROSAT or Suzaku) enhancements at the ends of observation segments may be due to solar X-rays scattered from the Earth's atmosphere or from aurorae. Lightcurves constructed in almost any broad band will be sufficient to detect these enhancements.

A lightcurve constructed in the 2.5–8.0 keV band (or any band where the SPC is strong compared to the source and the particle background) should be used to identify and remove time periods contaminated by soft protons. A lightcurve in the 0.45–0.75 keV band, particularly in conjunction with solar-wind data from the ACE or Wind missions, can be used to detect the presence of SWCX. If analysis depends, for example, on the O lines, and there is variation in this band, then there is little that can be done except remove time periods with elevated flux. Even then the resulting analysis will be suspect.

The instrument manual may suggest additional lightcurve filtering schemes.

Point-source removal: Removing point sources usually removes a significant source of noise at only a small expense of data from the diffuse emission of interest. In the absence of significant absorption along the line of sight, those sources contribute to the AGN background, so the strength of that background must be adjusted for the excluded sources using empirical AGN luminosity functions. Thus, it is best to remove the point sources to a consistent level to make that adjustment as straightforward as possible. For diffuse emission within the Galactic plane, the situation is both more simple and more complicated; the AGN are heavily absorbed by the Galactic disk, but there is substantial contribution from Galactic sources whose luminosity function is poorly understood and whose absorption will depend upon their spatial distribution. Stars have soft thermal spectra and are numerous but faint, contributing only below 1.5 keV for unabsorbed lines of sight. Galactic binaries, typically at large distances and strongly absorbed, have power-law-like spectra that can easily be distinguished from the thermal spectra of most diffuse emission.

8.4 Spectral analysis

Setting the source and background regions is a question of scientific need, instrumental necessity, and personal taste; extracting the spectra from those regions, a question of preferred analysis package; and spectral fitting, a question of experience. The important point to be kept in mind is that the only background that can be directly subtracted from the observed spectrum (source plus all backgrounds) is the particle background; the remainder of the backgrounds must be modeled. In all of the following examples, it is assumed that the particle background spectrum has been constructed and subtracted.

The soft Galactic-halo emission: This is one of the most difficult observations to analyze as the emission fills the entire FOV and must somehow be separated from the soft LHB emission which also fills the FOV. The spectrum should be extracted from the entire FOV after removing point sources. The soft-proton contamination and the unresolved AGN can be fit simultaneously with the models for the LHB, the halo, and the extragalactic background. Since the emission is very low energy, the RASS data should be used to constrain the fits. The SWCX is an added complication that should be fit, but no adequate models exist.

A pulsar wind nebula (PWN) in an SNR: Supernova remnants (SNR) typically occur in the Galactic plane, so the diffuse Galactic foreground may be difficult to remove. Conversely, that emission is only important below $\sim$1 keV, where the SNR and PWN emission may be strongly absorbed. Thus, the Galactic foreground might be removed simply by limiting the energy range of the fit. The SNR produces a strongly varying (both spatially and spectrally) background that is hard to characterize for the position of the PWN. The PWN has a power-law signature (like the extragalactic background) so the power-law signatures in parts of the field far from the PWN must be characterized. All the components must be fit simultaneously with the PWN. For the PWN in the SNR W44, the background SNR was fit for many different regions around the PWN and it was found that they could all be fit by a simple combination of thermal models with constant temperatures but variable normalizations (Petre *et al.*, 2002). This background model was used in the PWN region, and only the normalizations were varied while fitting the spectral shape of the PWN.

Clusters of galaxies: The most interesting part of a cluster is often the outer parts, which have surface brightnesses comparable to or below that of the Galactic foreground. As there is considerable spectral overlap, this is a particular problem for clusters that (nearly) fill the FOV. In addition, multiple annuli must be fit simultaneously in order to determine the temperature profile. Assuming that the Galactic foreground is constant across the cluster, the same model with the same parameters can be used for each annulus; the effect of vignetting is included in the responses. For XMM–Newton, the soft-proton contamination has the same spectral shape across most of the detector so all annuli should have the same spectral-shape parameters, but the normalizations should vary. Finally, the source spectrum will be different in each annulus, and XMM–Newton's large PSF will scatter flux from one annulus to another. Setting up such a fit can be quite complex; recent work with XMM–Newton data required 546 fit parameters, cross-linked in complicated ways, to fit 10 annuli (Snowden *et al.*, 2008). However, that complexity is necessary to avoid systematic problems introduced by overly simplistic background models.

8.5 Image analysis

It may seem strange to delay image analysis until after the (seemingly) more complex spectral analysis. However, given the variation of the response with position and energy, the image analysis should always be shaped by an understanding of the source and background spectra. An important point to keep in mind is that the broader the bandpass, the more difficult the background removal and subsequent analysis.

Building the right effective area map: In the most general analysis, the fluxed image (counts/cm^2/s/pixel) is created by dividing a raw image (counts/pixel) by the effective area (cm^2) and the exposure time. The effective-area map, $EA(i, j, E)$ is the equivalent of a flat-field or instrumental response map, and is a function of the pixel position, (i, j), and energy, E (see Section 4.5.1). A monochromatic effective-area map is calculated for a single energy. Most images are constructed in a broad energy band, so an emission-weighted effective-area map should be employed:

$$R(i, j) = \sum_E EA(i, j, E)S(E)/ \sum_E S(E) \qquad (8.2)$$

where $S(E)$ is the spectrum of the emission and the sum is over the bandpass.[5] Clearly, analysis becomes much more difficult when the source spectrum varies with position!

Building background maps from background spectra: Since the background components have different spectral shapes, and thus different distributions across the detector (due the energy dependence of the response) it is usually a good idea to remove all of the backgrounds before dividing the raw image by the effective-area map. At the very least, the particle background image, $PB(i, j)$, and the soft-proton image $SP(i, j)$ should be subtracted, as these images will have very different distributions compared to the response

---

[5] The messy looking mathematics are not required to understand what is going on here, but absolutely essential when trying to figure out why the image does not look right. This formulation is very useful for discovering missing normalizations!

to X-ray photons. If $R_b(i, j)$ is the effective-area map created from/for the background spectrum, $S_b(E)$, then the background image is:

$$C_b(i, j) = R_b(i, j) \sum_E S_b(E)t \tag{8.3}$$

where t is the exposure time.

Subtracting the backgrounds: If $I(i, j)$ is the raw image, then the flux image of the source is the raw-count image from which all of the backgrounds are subtracted, which is then divided by the effective-area map appropriate for the source spectrum:

$$= \frac{1}{R_S(i, j)t} \left[I(i, j) - PB(i, j) - SP(i, j) - \sum_N \left[R_n(i, j) \sum_E S_n(E)t \right] \right] \tag{8.4}$$

where R_S is the effective area map for the source spectrum, R_n are the effective area maps for the X-ray photon backgrounds, and the sum N is over all of the X-ray-photon backgrounds, S_n (Galactic foreground, extragalactic background, exospheric emission, etc.). The Galactic foreground and extragalactic background should be relatively flat across the FOV so whether or not these backgrounds should be subtracted before the division may not be a clear decision even in the usual case that the source and background spectra are very different. If the spatial variation of the response to the background is very similar to the spatial variation of the response to the source, that is, for all pixels (i, j)

$$\left[\frac{R_S(i, j)}{R_B(i, j)} - \frac{\overline{R}_S}{\overline{R}_B} \right] C_B \ll C_S \tag{8.5}$$

where C_B and C_S are the background and source surface brightnesses and the means are constructed over the region of interest, then a flux image created by dividing the raw image by the effective-area map and *then* subtracting the backgrounds,

$$\frac{I(i, j) - [PB(i, j) + SP(i, j)]}{R_S(i, j)t} - \sum_N \left[\frac{R_n}{R_S} \sum_E S_n(E) \right] \tag{8.6}$$

will be sufficient for analyses in which small systematic uncertainties are not important, such as a bright SNR seen against the Galactic plane. That is, if the difference in effective area maps is small and the background is much fainter than the source, then subtracting a constant background after the division is fine.

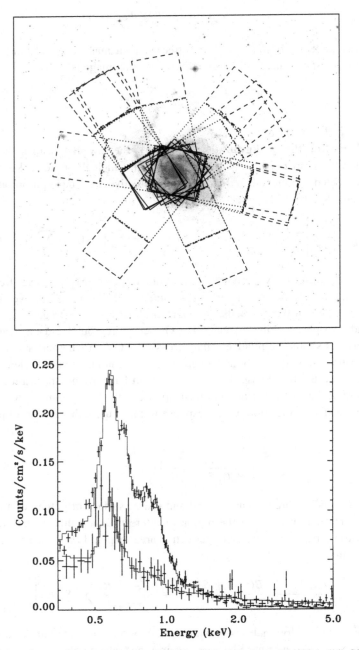

Fig. 8.6 [Top] An optical image of M101 with the locations of the S3 (solid), S2 (dotted), and S1 chips (dashed) marked. In this case, there were multiple exposures at different positions and roll angles. [Bottom] A spectrum of M101 (upper line) and the Galactic foreground (lower line)

A more simple background construction: This analysis assumes that the spectral shape of the background has been fit (or is otherwise known). Therefore, the background subtraction will only be as good as the spectral fit; if the fit is poor, the background subtraction will be as well. One way around this problem is to use the poor spectral fit to get a (roughly) correct effective-area map, and then to use the measured number of background counts, $\overline{B}$, in some region (denoted by primes) to get:

$$C_B(i, j) = \left[\frac{\overline{B}}{\sum R_B(i', j')} \right] R_B(i, j) \qquad (8.7)$$

Example analysis of a nearby galaxy: Figure 8.6 demonstrates one of the more challenging analysis problems. The M101 galaxy is covered by a sequence of ACIS exposures for which some chips (S3, covering the center of the galaxy, and S1, at the periphery) have strong soft response, while others (S2, covering the outer arms) do not. The S3 chip has very little area off the galaxy, and that area is contaminated with Galactic emission. The background is well covered by the S1 chip, but that chip has a very different response. In this case, the background spectrum from the S1 chip (and possibly that portion of the S3 chip not obviously contaminated by the galaxy) was fit in order to determine the Galactic foreground and AGN background.

The spectrum of the galaxy (Figure 8.6 [bottom]), shows that the bulk of the emission is below ~ 1.25 keV. For scientific reasons, the primary interest was in two narrow-band images with $0.35 < E < 0.7$ and $0.7 < E < 1.25$ keV. These bandpasses are sufficiently narrow that a monochromatic effective-area map could have been used for each.[6] However, since the response at these energies is varying rapidly with energy, determining which energy to choose could be difficult. Since, for Chandra data, an emission-weighted effective area map can be made easily using CIAO, it was easier to make the emission-weighted map than to determine the energy for a monochromatic map. From the spectral fit to the Galactic background the count rate in each bandpass was determined, that rate was multiplied by the background effective-area map to create an image of the background, which was then subtracted from the raw image. A similar method was used to construct and remove the extragalactic background. In this case, the problem becomes a bit more difficult as the absorption of the extragalactic background by the galaxy changes the spectral shape as a function of position. (Here the spectral shape of the background should be taken from the place where the background has the strongest signal.)

---

[6] Remember that if a monochromatic effective-area map at energy E_0 with background values obtained from fitting is used, $C_B(i, j) = R_{E_0}(i, j) \left[\frac{\sum \mathcal{R}(E) S_B(E)}{R(E_0) \sum S(E)} \right] \sum S_B(E)$ (where $\mathcal{R}$ is the response used for the fit) must be renormalized.

Those who already have experience with this type of analysis may think that this prescription is unnecessarily complex, and in some situations it is. Often monochromatic effective-area maps *can* be used and the difference in spectral shape among the various components ignored. However, since the Galactic emission is relatively soft, and this is the spectral region in which current instruments have responses that vary strongly with energy, this more complex prescription avoids unnecessary errors. Due to the optical design, the effective response usually changes most rapidly near the edge of the detector. Imagine searching for a faint extended Galactic halo when the galaxy has been placed on the optical axis. Clearly, a very small difference between the true and assumed shape of the effective-area profile could produce a spurious signal that mimicks that expected for a halo.

8.6 Mosaics

Diffuse emission is frequently larger than the FOV so it is often the case that multiple images must be mosaiced to cover an extended object. Although a number of mission-specific programs have been, or are being, developed for this task, there is no universally applicable software. In general, the operation has two steps; to recast the raw-count images, effective-area maps, and background images for each exposure into a common coordinate system, and then to combine all of the recast images in the same way that individual images would be combined. Mosaicking is a severe test of the background removal; any error in the strength or distribution of the backgrounds will produce obvious artifacts in the mosaic.

In the low-count-rate regime, where there is significantly less than one count per pixel, division of the raw count image by the exposure time will greatly enhance the values of the individual pixels containing non-zero counts in the lower-exposure regions compared to higher-exposure regions. The resulting mosaic is not only esthetically unpleasing, but the eye is incapable of interpreting properly the true variation in surface brightness. For this reason, the mosaic is generally smoothed over areas large enough to include many counts. As that area may vary strongly across the mosaic, *adaptive* smoothing is often appropriate.

Given that the different exposures may have been taken at very different times with different backgrounds and instrumental responses, it is often not appropriate to use mosaics for analysis; when possible, return to the original exposures for photometry or spectroscopy.

Appendix 1

X-ray lines and edges

The X-ray waveband contains atomic and ionic transitions for nearly all astrophysically abundant elements – with the notable exception of H and He. These arise primarily from transitions involving electrons in the $1s$ shell but for heavier elements (i.e. Fe, Ni), there are transitions involving higher shells as well. This appendix contains a short discussion of spectroscopic notation combined with information on a selection of particularly strong transitions, including those from hydrogen-like and helium-like ions, Fe XVII–Fe XXIV, as well as fluorescent transitions from neutral atoms and ionization edges for all of the abundant elements. More information about atomic data useful for X-ray astronomy can be found at http://www.atomdb.org.

A1.1 Spectroscopic notation

A complete discussion of spectroscopic notation is beyond the scope of this handbook; we suggest the short but highly informative text by Herzberg (1945) for a more detailed review; another useful source is the *X-ray Data Booklet* published by the Center for X-ray Optics and Advanced Light Source at Lawrence Berkeley National Laboratory (http://xdb.lbl.gov). It should be noted that X-ray astronomy is rife with poorly used spectroscopic terminology, so following the form used by an earlier refereed paper does not guarantee proper usage.

Electron shells: Atomic theory states that the electrons in an atom or ion are bound into "shells" with principal quantum number n starting at 1 and increasing indefinitely. The shells with quantum number $n = 1, 2, 3$, and 4 are typically referred to as the K, L, M, and N shells respectively. X-ray astronomy typically involves transitions to the K or L shells, although Fe M-shell transitions (i.e. $\Delta n = n' \rightarrow 3$) can generate photons near 0.1 keV. This nomenclature can be further divided into subshells that also give the orbital

Table A1.1   *Hydrogen-like Lyα lines from Johnson and Soff (1985). Lyβ*
lines from Erickson (1977)

Spectral line	Lyα$_1$/ Å	Lyα$_1$/ keV	Lyα$_2$/ Å	Lyα$_2$/ keV	Lyβ$_1$/ Å	Lyβ$_1$/ keV	Lyβ$_2$/ Å	Lyβ$_2$/ keV
C VI	33.734	0.3675	33.740	0.3675	28.465	0.436	28.466	0.436
N VII	24.779	0.5004	24.785	0.5002	20.909	0.593	20.911	0.593
O VIII	18.967	0.6537	18.973	0.6535	16.006	0.775	16.007	0.775
Ne X	12.132	1.0220	12.137	1.0215	10.238	1.211	10.240	1.211
Na XI	10.023	1.2370	10.029	1.2363	8.459	1.466	8.460	1.465
Mg XII	8.419	1.4726	8.425	1.4717	7.106	1.745	7.107	1.745
Al XIII	7.171	1.7290	7.176	1.7277	6.089	2.036	6.091	2.036
Si XIV	6.180	2.0061	6.186	2.0043	5.217	2.377	5.218	2.376
S XVI	4.727	2.6227	4.733	2.6197	3.991	3.107	3.992	3.106
Ar XVIII	3.731	3.3230	3.737	3.3182	3.150	3.936	3.151	3.934
Ca XX	3.018	4.1075	3.024	4.1001	2.549	4.864	2.550	4.862
Fe XXVI	1.778	6.9732	1.783	6.9520	1.502	8.252	1.504	8.246
Ni XXVIII	1.530	8.1017	1.536	8.0731	1.293	9.586	1.295	9.577

angular momentum of the electrons. This is given as an integer subscript to the
shell letter, but for historical reasons the value is incremented by 1. Thus the
L_1 shell refers to electrons with principal quantum number $n = 2$ and orbital
angular momentum $l = 0$ (also known as $2s$ shell) while the M_3 shell contains
electrons with $n = 3$ and $l = 2$, or the $3d$ shell. Somewhat confusingly, some
authors use Roman numerals (e.g. I, II, III) instead of Arabic for the subscript,
although this is less common.

A1.1.1 Describing line emission

Ideally, every emission or absorption line would have a single unambiguous label that fully describes the transition. For simple systems that are well
described by LS coupling, such as helium-like ions, standard notation such as
$1snl \, ^{(2S+1)}L_J$ can be used. This notation describes a system with one electron in
the $1s$ shell, one in a shell with principal quantum number n and angular momentum l; the whole system has total spin S, orbital angular momentum L, and
total angular momentum J. Unfortunately, the realities of quantum mechanics,
especially for complex ions with multiple electrons, makes this goal impractical. In practice, a range of different terms are used depending upon historical
precedent, the bandpass at issue, and the taste or training of the author.

Before listing individual ions, it should be noted that there is a distinct
difference in meaning when using Roman numerals versus Arabic superscripts.
O^{+6} refers to an ion of oxygen with two bound electrons and a positive charge of

Table A1.2   *Helium-like R, I_2, I_1, F lines from Drake (1988)*

Spectral line	R[w]/ Å	R[w]/ keV	I_2 [x]/ Å	I_2 [x]/ keV	I_1 [y]/ Å	I_1 [y]/ keV	F [z]/ Å	F [z]/ keV
C V	40.268	0.3079	40.729	0.3044	40.73	0.3044	41.472	0.2990
N VI	28.787	0.4307	29.082	0.4263	29.08	0.4264	29.534	0.4198
O VII	21.602	0.5739	21.801	0.5687	21.804	0.5686	22.101	0.5610
Ne IX	13.447	0.9220	13.550	0.9150	13.553	0.9148	13.698	0.9051
Na X	11.003	1.1268	11.080	1.1190	11.083	1.1187	11.192	1.1078
Mg XI	9.169	1.3522	9.228	1.3435	9.231	1.3431	9.314	1.3311
Al XII	7.757	1.5983	7.804	1.5888	7.807	1.5881	7.872	1.5750
Si XIII	6.648	1.8650	6.685	1.8547	6.688	1.8538	6.740	1.8395
S XV	5.039	2.4606	5.063	2.4488	5.066	2.4471	5.102	2.4303
Ar XVII	3.949	3.1396	3.966	3.1262	3.969	3.1238	3.994	3.1043
Ca XIX	3.177	3.9025	3.189	3.8879	3.193	3.8830	3.211	3.8612
Fe XXV	1.851	6.7000	1.855	6.6823	1.859	6.6676	1.868	6.6365
Ni XXVII	1.588	7.8053	1.592	7.7865	1.597	7.7660	1.604	7.7316

6, while O VII is a spectroscopic notation describing an emission or absorption line arising from electronic transitions within an O^{+6} ion. Thus, O VII describes a *spectrum of lines* or possibly a *photon*, while O^{+6} describes an *ion*; the two are not equivalent. For this reason, although the fully stripped ion O^{+8} will occur in a sufficiently hot plasma, there is no reason ever to use the notation O IX, since no electronic transition lines can occur in an ion with no electrons. O^{+8} can, of course, radiatively recombine to form O^{+7}, but the emission process generates a continuum, not an emission line. Although the written distinction is clear, confusion can arise when speaking and so astronomers are advised to say O^{+6} as "oxygen plus six" while O VII is "oxygen seven," especially when speaking to atomic physicists or chemists.

Hydrogen-like (i.e. ions with a single electron) line emission follows the nomenclature used for hydrogen line emission itself. Transitions ending on the $n = 1, 2$, and 3 shells are described as Lyman, Balmer (or Hydrogen), and Paschen series (usually shortened to Ly, H, or Pa). Within each series, Greek letters are then used to indicate the delta of the principal quantum number for the transition, so that α, β, γ refer to $\Delta n = 1, 2, 3$, respectively. Thus Lyα is the transition from $n = 2 \rightarrow 1$ in a hydrogen-like ion, while Hγ is the transition from $n = 5 \rightarrow 2$ and Paβ goes from $n = 5 \rightarrow 3$. These transitions are all doublets and so can be further subdivided with subscripts. These are defined so that the subscript 1 indicates the transition line with the larger total angular momentum while subscript 2 indicates the smaller angular momentum. Thus, Lyα_1 is the transition $2p^2P_{3/2} \rightarrow 1s^2S_{1/2}$, while Ly$\beta_2$ is the transition $3p^2P_{1/2} \rightarrow 1s^2S_{1/2}$.

Table A1.3   *Fluorescent emission lines (in keV), from*
Krause and Oliver (1979)

Element	$K\alpha_1$	$K\alpha_2$	$K\beta_1$	$L\alpha$	$L\beta$
C	0.277				
N	0.3924				
O	0.5249				
Ne	0.8486	0.8486			
Na	1.0410	1.0410	1.0711		
Mg	1.2536	1.2536	1.3022		
Al	1.4867	1.4863	1.5575		
Si	1.7400	1.7394	1.8359		
S	2.3078	2.3066	2.4640		
Ar	2.9577	2.9556	3.1905		
Ca	3.6917	3.6881	4.0127	0.3413	
Fe	6.4038	6.3908	6.4905	0.6374	0.6488
Ni	7.4782	7.4609	8.2647	0.8515	0.8588

Helium-like line emission can most simply be described using standard LS notation, i.e. $1s3p\ ^1P_1 \to 1s^2\ ^1S_0$, although there are special cases for transitions from $n = 2 \to 1$. These are listed in Table 5.2.

Ions with more electrons become progressively more problematic because the LS-term designation breaks down due to electron–electron interactions within the ion. A common solution to this problem is to use designations developed from experimental results. For example, labels for strong lines from Fe XVII through Fe XXIV (see Table A1.4) can be found in Brown *et al.* (1998), although these are not universally used.

Fluorescent emission lines can arise when an atom or ion has an inner-shell electron removed, possibly due to photoionization, electron collision, or nuclear transition. After the event, the ion is highly unstable and will rapidly fill the electron "hole" with an electron from a less-bound shell, releasing energy that may ionize the ion further or emerge as a photon. The ionization case is generally termed the "Auger" process, although technically this only applies if the electron "hole", the electron filling the hole, and the ionized electron all come from distinct shells, such as an L-shell electron filling a K-shell hole and ionizing an M-shell electron in the process; this is often written as a KLM transition. If the vacancy and the transitioning electrons are from the same shell

Spectral line	λ(Å)	E (keV)	Label[a]	Transition
Fe XXIV	10.619	1.1676		$3p\ ^2P_{3/2} \rightarrow$ ground state
Fe XXIII	10.981	1.1291	Be9	$2s3p\ ^1P_1 \rightarrow$ ground state
Fe XXIII	11.019	1.1252	Be8	$2s3p\ ^3P_1 \rightarrow$ ground state
Fe XXIV	11.029	1.1242	Li4	$3d\ ^2D_{3/2} \rightarrow 2p\ ^2P_{1/2}$
Fe XXIV	11.176	1.1094	Li3	$3d\ ^2D_{5/2} \rightarrow 2p\ ^2P_{3/2}$
Fe XVII	11.254	1.1017	5D	$2p^55d\ ^3D_1 \rightarrow$ ground state
Fe XXIV	11.432	1.0845	Li1	$3s\ ^2S_{1/2} \rightarrow 2p\ ^2P_{3/2}$
Fe XXIII	11.736	1.0564	Be2	$2s3d\ ^1D_2 \rightarrow 2s2p\ ^1P_1$
Fe XXII	11.770	1.0534	B13	$2s^23d\ ^2D_{3/2} \rightarrow$ ground state
Fe XVII	12.124	1.0226	4C	$2p^54d\ ^1P_1 \rightarrow$ ground state
Fe XXIII	12.161	1.0195	Be1	$2s3s\ ^1S_0 \rightarrow 2s2p\ ^1P_1$
Fe XVII	12.266	1.0108	4D	$2p^54d\ ^3D_1 \rightarrow$ ground state
Fe XXI	12.284	1.0093	C10	$2s^22p3d\ ^3D_1 \rightarrow$ ground state
Fe XXI	12.393	1.0004	C8	$2s^22p3d\ ^3D_1 \rightarrow 2s^22p^2\ ^3P_1$
Fe XXII	12.754	0.9721	B4	$2s2p_{1/2}3s \rightarrow 2s2p^2\ ^2D_{3/2}$
Fe XX	12.846	0.9652	N31	$2s^22p_{1/2}2p_{3/2}3d_{3/2} \rightarrow$ ground state
Fe XX	12.864	0.9638	N31	$2s^22p_{1/2}2p_{3/2}3d_{5/2} \rightarrow$ ground state
Fe XXI	12.822	0.9670	C4	$2s2p_{1/2}2p_{3/2}3d_{5/2} \rightarrow 2s2p^3\ ^3D_1$
Fe XX	12.824	0.9668	N31	$2s^22p_{1/2}2p_{3/2}3d_{3/2} \rightarrow$ ground state
Fe XX	12.912	0.9602	N30	$2s^22p_{1/2}2p_{3/2}3d_{5/2} \rightarrow$ ground state
Fe XX	12.965	0.9563	N29	$2s^22p_{1/2}2p_{3/2}3d_{3/2} \rightarrow$ ground state
Fe XX	13.061	0.9493	N26	$2s^22p_{1/2}^23d_{5/2} \rightarrow$ ground state
Fe XIX	13.462	0.9210	O26	$2s^22p^3(^2D)3d\ ^3S_1 \rightarrow$ ground state
Fe XIX	13.497	0.9186	O25	$2s^22p^3(^2D)3d\ ^3P_2 \rightarrow$ ground state
Fe XXI	13.507	0.9179	C3	$2s2p_{1/2}^23s \rightarrow 2s2p^3\ ^3D_1$
Fe XIX	13.518	0.9172	O24	$2s^22p^3(^2D)3d\ ^3D_3 \rightarrow$ ground state
Fe XIX	13.795	0.8988	O19	$2s^22p^3(^4S)3d\ ^3D_3 \rightarrow$ ground state
Fe XVII	13.825	0.8968	3A	$2s2p^63p\ ^1P_1 \rightarrow$ ground state
Fe XXI	14.008	0.8851	C1	$2s^22p_{1/2}3p_{1/2} \rightarrow 2s2p^3\ ^3D_1$
Fe XVIII	14.208	0.8726	F20	$2s^22p^4(^1D)3d\ ^2D_{5/2} \rightarrow$ ground state
Fe XVIII	14.208	0.8726	F20	$2s^22p^4(^1D)3d\ ^2D_{3/2} \rightarrow$ ground state
Fe XVIII	14.256	0.8697	F19	$2s^22p^4(^1D)3d\ ^2S_{1/2} \rightarrow$ ground state
Fe XX	14.267	0.8690	N9	$2s2p_{1/2}^22p_{3/2}3s \rightarrow 2s2p^4\ ^4P_{5/2}$
Fe XVIII	14.373	0.8626	F17	$2s^22p^4(^3P)3d\ ^2D_{5/2} \rightarrow$ ground state
Fe XVIII	14.534	0.8531	F15	$2s^22p^4(^3P)3d\ ^2F_{5/2} \rightarrow$ ground state
Fe XIX	14.664	0.8455	O13	$2s^22p^3(^2D)3s\ ^3D_3 \rightarrow$ ground state
Fe XVII	15.014	0.8258	3C	$2p^53d\ ^1P_1 \rightarrow$ ground state
Fe XIX	15.079	0.8222	O8	$2s^22p^3(^4S)3s\ ^5S_2 \rightarrow$ ground state
Fe XIX	15.198	0.8158	O5	$2s2p_{1/2}^22p_{3/2}^23s \rightarrow 2s2p^5\ ^3P_2$
Fe XVII	15.261	0.8124	3D	$2p^53d\ ^3D_1 \rightarrow$ ground state
Fe XVII	15.453	0.8023	3E	$2p^53d\ ^3P_1 \rightarrow$ ground state
Fe XVIII	15.625	0.7935	F11	$2s^22p^4(^1D)3s\ ^2D_{5/2} \rightarrow$ ground state
Fe XVIII	15.824	0.7835	F9	$2s^22p^4(^3P)3s\ ^4P_{3/2} \rightarrow$ ground state
Fe XVIII	16.004	0.7747	F6	$2s^22p^4(^3P)3s\ ^2P_{3/2} \rightarrow$ ground state
Fe XVIII	16.071	0.7715	F4	$2s^22p^4(^3P)3s\ ^4P_{5/2} \rightarrow$ ground state
Fe XIX	16.110	0.7696	O2	$2s^22p_{1/2}2p_{3/2}^23p_{1/2} \rightarrow 2s2p^5\ ^3P_2$
Fe XVIII	16.159	0.7673	F3	$2s2p^53s\ ^2P_{3/2} \rightarrow 2s2p^6\ ^2S_{1/2}$
Fe XVII	16.780	0.7389	3F	$2p^53s\ ^1P_1 \rightarrow$ ground state
Fe XVII	17.051	0.7271	3G	$2p^53s\ ^3P_1 \rightarrow$ ground state
Fe XVII	17.096	0.7252	M2	$2p^53s\ ^3P_2 \rightarrow$ ground state
Fe XVIII	17.623	0.7035	F1	$2s^22p^43p\ ^2P_{3/2} \rightarrow 2s2p^6\ ^2S_{1/2}$

[a]From Brown *et al.* (1998).

Table A1.5 *Ionization edges (in eV), from Verner et al. (1996)*

Ion	C	N	O	Ne	Na	Mg	Al	Si	S	Ar	Ca	Fe
I	11.26	14.53	13.62	21.56	5.139	7.646	5.986	8.152	10.36	15.76	6.113	7.902
II	24.38	29.60	35.12	40.96	47.29	15.04	18.83	16.35	23.33	27.63	11.87	16.19
III	47.89	47.45	54.94	63.46	71.62	80.14	28.45	33.49	34.83	40.74	50.91	30.65
IV	64.49	77.47	77.41	97.12	98.92	109.3	120.0	45.14	47.31	59.81	67.27	54.80
V	392.1	97.89	113.9	12.62	138.4	141.3	153.8	166.8	72.68	75.02	84.51	75.01
VI	490.0	552.1	138.1	157.9	172.2	186.5	190.5	205.1	88.05	91.01	108.8	99.06
VII		667.1	739.3	207.3	208.5	224.9	241.4	246.5	280.9	124.3	127.2	125.0
VIII			871.4	239.1	264.2	266.0	284.6	303.2	328.2	143.5	147.2	151.1
IX				1196	299.9	328.2	330.1	351.1	379.1	422.5	188.3	233.6
X				1362	1465	367.5	399.4	401.4	447.1	478.7	211.3	262.1
XI					1649	1762	442.1	476.1	504.8	539.0	591.9	290.2
XII						1963	2086	523.5	564.7	618.3	657.2	330.8
XIII							2304	2438	651.7	686.1	726.7	361.0
XIV								2673	707.2	755.8	817.7	392.2
XV									3224	854.8	894.6	457.0
XVI									3494	918.0	974.5	489.3
XVII										4121	1087	1262
XVIII										4426	1157	1358
XIX											5129	1456
XX											5470	1582
XXI												1689
XXII												1799
XXIII												1950
XXIV												2046
XXV												8829
XXVI												9278

(e.g. an LLM transition), the process is properly termed a "Coster–Kroenig" transition, although this nuance is often overlooked.

Alternatively, the unstable ion may radiatively stabilize by emitting a "fluorescent" photon. The relative probability of ionization vs. fluorescence depends primarily upon the nuclear charge; low-Z elements overwhelmingly tend to ionize, while iron with a K-shell hole has a $\sim 30\%$ chance of emitting a photon and elements much heavier than iron strongly tend towards fluorescence. Fluorescence lines are described first by the shell of the electron hole, so an "Fe K" line refers to a transition ending with an electron moving to the K-shell. Much like hydrogen-like ions, this can be refined with a Greek letter such as α or β to indicate the Δn of the transition, so that a Kα transition arises from a $n = 2 \to 1$ transition and Kβ from $n = 3 \to 1$. Finally, a subscript to the Greek letter further subdivides the transitions, although the author is not aware of any general rule.

For any given ion with at least one electron, the ionization "edge" refers to the least amount of energy required (either collisional or radiative) to remove an electron from the ion. Table A1.5 lists these values for all ionization stages of the astrophysically abundant elements in the gas phase. However, these values will shift when the element is in a solid or molecule.

Appendix 2

Conversion tables

RANDALL K. SMITH

A2.1 Useful equations

We include here equations to convert from typical optical/UV parameters, such as extinction, to the hydrogen column densities used in X-ray astronomy, as well as equations for thermal broadening and conversion between flux units.

A2.1.1 Absorption

X-ray absorption is strongly energy-dependent (scaling roughly as E^{-3}), with the primary sources of absorption changing as a function of energy. In general, He atoms dominate at lower energies with abundant metals such as C and O becoming significant around 0.3–0.5 keV. Despite the relative insignificance of hydrogen, the total absorption is usually given in terms of a hydrogen cross-section N_H. Table A2.1 shows the empirical conversions with the more typical values A_V and E_{B-V} used in the optical, UV, and IR wavebands.

A2.1.2 Line broadening

Line broadening can be caused by strong magnetic or electric fields, or by intrinsic atomic or thermal effects. In the X-ray-emitting plasmas, the dominant term is typically due to thermal Doppler shifts. The relevant equations are shown in Table A2.2.

A2.2 Useful astrophysical values

We include a selection of useful constants, conversion factors, and measured values that are often useful in astrophysical research. The abundances of common elements are given in Table A2.3, taken from Asplund *et al.* (2005). However, the entire issue of cosmic and solar abundances has been changing, so the

Table A2.1 *Absorption and extinction*

N_H/A_V	=	1.9×10^{21}	atoms/cm²/mag
N_H/E_{B-V}	=	$(5.9 \pm 1.6) \times 10^{21}$	atoms/cm²/mag

From Seward (2000).

Table A2.2 *Thermal broadening and Maxwell–Boltzmann (MB) equation*

MB Eq. for velocity	$f(\vec{v})d\vec{v} = \left(\frac{m}{2\pi k_B T}\right)^{3/2} \exp\left(\frac{-m\vec{v}^2}{2k_B T}\right)dv$
MB Eq. for speed	$f(v)dv = 4\pi \left(\frac{m}{2\pi k_B T}\right)^{3/2} v^2 \exp\left(\frac{-mv^2}{2k_B T}\right)dv$
Line broadening	$\frac{\Delta\lambda}{\lambda} = 7.16 \times 10^{-7}\sqrt{\frac{T_K}{M_{amu}}}$
Resolution	$\frac{\lambda}{\Delta\lambda} = \frac{E}{\Delta E} \equiv R = 1.4 \times 10^6 \sqrt{\frac{M_{amu}}{T_K}}$
...	or $R = 410\sqrt{\frac{M_{amu}}{T_{keV}}}$

Table A2.3 *Abundances of common elements, from Asplund et al. (2005)*

Element	Solar photosphere	Meteorites
H	12.00	8.25 ± 0.05
He	$[10.93 \pm 0.01]^1$	1.29
C	8.39 ± 0.05	7.40 ± 0.06
N	7.78 ± 0.06	6.25 ± 0.07
O	8.66 ± 0.05	8.39 ± 0.02
F	4.56 ± 0.30	4.43 ± 0.06
Ne	$[7.84 \pm 0.06]^1$	−1.06
Na	6.17 ± 0.04	6.27 ± 0.03
Mg	7.53 ± 0.09	7.53 ± 0.03
Al	6.37 ± 0.06	6.43 ± 0.02
Si	7.51 ± 0.04	7.51 ± 0.02
P	5.36 ± 0.04	5.40 ± 0.04
S	7.14 ± 0.05	7.16 ± 0.04
Cl	5.50 ± 0.30	5.23 ± 0.06
Ar	$[6.18 \pm 0.08]^1$	−0.45
K	5.08 ± 0.07	5.06 ± 0.05
Ca	6.31 ± 0.04	6.29 ± 0.03
Ti	4.90 ± 0.06	4.89 ± 0.03
V	4.00 ± 0.02	3.97 ± 0.03
Cr	5.64 ± 0.10	5.63 ± 0.05
Mn	5.39 ± 0.03	5.47 ± 0.03
Fe	7.45 ± 0.05	7.45 ± 0.03
Co	4.92 ± 0.08	4.86 ± 0.03
Ni	6.23 ± 0.04	6.19 ± 0.03
Cu	4.21 ± 0.04	4.23 ± 0.06
Zn	4.60 ± 0.03	4.61 ± 0.04

[1] Indirect solar estimates marked with [..]

Table A2.4 *Useful physical constants and conversion factors*

Quantity	Symbol / Equation	Value	Units
Speed of light	c	2.99792458×10^8	m/s
Gravitational constant	G	6.673×10^{-11}	m^3/kg/s^2
Rydberg energy	$hcR_\infty = m_e c^2 \alpha^2/2$	13.60569	eV
Planck constant	h	6.626×10^{-27}	erg s
	$\hbar \equiv h/2\pi$	1.054572×10^{-27}	erg s
Electron mass	m_e	9.109389×10^{-28}	g
		510.999	keV/c^2
Proton mass	m_p	1.672622×10^{-28}	g
		938.272	MeV/c^2
Atomic mass unit	mass of C^{12}/12	1.66054×10^{-24}	g
		931.494	MeV/c^2
Electron charge	e	1.602177×10^{-19}	C
Electron-volt	eV	1.602177×10^{-12}	erg
Fine structure constant	$\alpha = e^2/(4\pi\epsilon_0\hbar c)$	1/137.03599	
Classical electron radius	$r_e = \alpha\hbar/(m_e c)$	2.817940×10^{-13}	cm
Thompson cross-section	$8\pi r_e^2/3$	6.652×10^{-25}	cm^{-2}
Electron Compton wavelength	$\hbar/m_e c = r_e\alpha^{-1}$	3.861593×10^{-11}	cm
Boltzmann constant	k_B	1.3807×10^{-23}	J/K^1
		8.6173×10^{-8}	keV/K^1
Stefan-Boltzmann constant	$\sigma = \pi^2 k_B^4/(60\hbar^3 c^2)$	5.670×10^{-8}	W/m^2/K^4
Wien's law constant	$b = \lambda_{max} T$	2.89777×10^{-3}	m K

Table A2.5 *Useful astronomical constants and conversion factors*

Quantity	Symbol	Value	Units
Solar mass	$M_\odot$	1.989×10^{33}	g
Solar radius	$R_\odot$	6.960×10^{10}	cm
Solar luminosity	$L_\odot$	3.826×10^{33}	erg s^{-1}
Earth mass	$M_\oplus$	5.976×10^{27}	gm
Earth radius	$M_\oplus$	6.378×10^3	km
Earth gravity	$g_\oplus$	9.807×10^2	cm s^{-2}
Astronomical unit	AU	1.496×10^{13}	cm
Light year	lyr	9.461×10^{17}	cm
Parsec	pc	3.086×10^{18}	cm
		3.262	lyr
Rayleigh	Ry	$(1/4\pi) \times 10^6$	ph cm^{-2}s^{-1}sr^{-1}
Jansky	Jy	10^{-26}	W m^{-2}Hz^{-1}
Steradian	sr	3.283×10^3	deg^2
		1.182×10^7	arcmin^2
		4.255×10^{10}	arcsec^2
Degree		1.745×10^{-2}	radian
Arcmin		2.909×10^{-4}	radian
Arcsec		4.848×10^{-6}	radian
Degree^2		3.046×10^{-4}	sr
Arcmin^2		8.462×10^{-8}	sr
Arcsec^2		2.350×10^{-11}	sr

Table A2.6 Energy-unit conversion

To → From ↓	λ (Å)	λ (μm)	λ (cm)	ν (Hz)	E (keV)	WN (cm^{-1})	E (erg)
λ (Å)	λ	$10^{-4}\lambda$	$10^{-8}\lambda$	$3.00 \times 10^{18}/\lambda$	$12.4/\lambda$	$10^{8}/\lambda$	$1.99 \times 10^{-8}/\lambda$
λ (μm)	$10^{4}\lambda$	λ	$10^{-4}\lambda$	$3.00 \times 10^{14}/\lambda$	$1.24 \times 10^{-3}/\lambda$	$10^{4}/\lambda$	$1.99 \times 10^{-12}/\lambda$
λ (cm)	$10^{8}\lambda$	$10^{4}\lambda$	λ	$3.00 \times 10^{10}/\lambda$	$1.24 \times 10^{-7}/\lambda$	$1/\lambda$	$1.99 \times 10^{-16}/\lambda$
ν (Hz)	$3.00 \times 10^{18}/\nu$	$3.00 \times 10^{14}/\nu$	$3.00 \times 10^{10}/\nu$	ν	$4.14 \times 10^{-18}\nu$	$3.34 \times 10^{-11}\nu$	$6.63 \times 10^{-27}\nu$
E (keV)	$12.4/E$	$1.24 \times 10^{-3}/E$	$1.24 \times 10^{-7}/E$	$2.42 \times 10^{17}E$	E	$8.07 \times 10^{6}E$	$1.60 \times 10^{-9}E$
WN (cm^{-1})	$10^{8}/WN$	$10^{4}/WN$	$1/WN$	$3.00 \times 10^{10}WN$	$1.24 \times 10^{-7}WN$	WN	$1.99 \times 10^{-16}WN$
E (erg)	$1.99 \times 10^{-8}/E$	$1.99 \times 10^{-12}/E$	$1.99 \times 10^{-16}/E$	$1.51 \times 10^{26}E$	$6.24 \times 10^{8}E$	$5.03 \times 10^{15}E$	E

[1] Thanks to Eureka Scientific, http://www.eurekasci.com, for the design of this table.

Table A2.7 Flux-density conversion

To → From ↓	S_ν (Jy)	f_E	f_λ	F_λ	F_ν
S_ν	S_ν	$1.51 \times 10^{3}\, S_\nu/E$	$1.51 \times 10^{3}\, S_\nu/\lambda$	$3.00 \times 10^{-5}\, S_\nu/\lambda^2$	$10^{-23}\, S_\nu$
f_E	$6.63 \times 10^{-4}\, E f_E$	f_E	$8.07 \times 10^{-2}\, E^2 f_E$	$1.29 \times 10^{-10}\, E^3 f_E$	$6.63 \times 10^{-27}\, E f_E$
f_λ	$6.63 \times 10^{-4}\, \lambda f_\lambda$	$8.07 \times 10^{-2}\, \lambda^2 f_\lambda$	f_λ	$1.99 \times 10^{-8}\, f_\lambda/\lambda$	$6.63 \times 10^{-27}\, \lambda f_\lambda$
F_λ	$3.34 \times 10^{4}\, \lambda^2 F_\lambda$	$4.06 \times 10^{6}\, \lambda^3 F_\lambda$	$5.03 \times 10^{7}\, \lambda F_\lambda$	F_λ	$3.34 \times 10^{-19}\, \lambda^2 F_\lambda$
F_ν	$10^{23}\, F_\nu$	$1.51 \times 10^{26}\, F_\nu/E$	$1.51 \times 10^{26}\, F_\nu/\lambda$	$3.00 \times 10^{18}\, F_\nu/\lambda^2$	F_ν

[1] Thanks to Eureka Scientific, http://www.eurekasci.com, for the design of this table.

question of which abundances are "preferred" remains unsettled. As a result, any paper using "solar" or "cosmic" abundances *must* reference what is used as the standard set. Common astronomical notation is used for the table, so that all abundances are taken relative to $\log H = 12.00$. Thus, the solar abundance of iron, relative to hydrogen, is $Fe/H = 10^{7.45-12.00} = 2.818 \times 10^{-5}$.

We present selected standard physical and astronomical constants and conversion factors in Tables A2.4 and A.2.5, respectively. Table A2.6 provides the values needed to convert between different wavebands, such as from wavelengths in microns to energies in kiloelectron-volts or wavenumbers in inverse centimeters. Table A.2.7 provides a similar set of conversions for flux units, such as S_ν (Jy), f_E (ph/cm^2/s/keV), f_λ (ph/cm^2/s/keV), F_λ (ergs/cm^2/s/Å), and f_ν (ergs/cm^2/s/Hz). In Table A2.7, values of E are assumed to be in kiloelectron-volts and λ in Ångstroms.

Appendix 3

Typical X-ray sources

RANDALL K. SMITH

This section is included as an aid to a beginner in X-ray astronomy who wishes to start the learning process by using a "good" source – one where adequate data can be guaranteed and no unusual circumstances make analysis difficult. For example, although Sco X-1 was the first X-ray source beyond the Solar System ever detected, it is so bright that it can be observed with modern detectors only in extremely unusual modes, making it a poor choice for today's beginner. The sources listed here have been regularly observed by numerous satellites in normal modes of operation and should provide good "test" cases for beginners. That said, there is nothing stopping observers from requesting observations even of common sources in unusual modes, so care should be exercised when selecting an observation.

A3.1 Point sources

Although they may have some intrinsic extent, the sources in Table A3.1 are all point sources as far as past and current X-ray telescopes are concerned. Observations of these sources may or may not involve gratings; this must be determined on an observation-by-observation basis.

A3.2 Diffuse sources

All of the sources in Table A3.2 are diffuse sources of varying extent. Some (such as the Cygnus Loop) will fill the FOV of any X-ray detector, while others (e.g. Cas A) will generally fit within the FOV of most instruments. Sources within the Solar System, such as Jupiter, move too rapidly for X-ray satellites to track them. In most cases, observations are done as a series of fixed pointings,

Table A3.1   *Common X-ray point sources, by source type*

Category	Sub-category	Source name
Active galactic nuclei	High-redshift ($z > 3$)	Q0420-388, GB1508+5714
	Mid-redshift ($z \sim 1$)	3C186, PKS1127-145
	Low-redshift ($z \approx 0.05$)	3C273
	Blazar	3C279
	BL Lac	PKS2155-304
	Seyfert Type I	NGC 5548
	Seyfert Type II	Mkn 348, NGC 1068
X-ray binaries	Low-mass	Aql X-1
	High-mass	X Per, Cyg X-1
	Eclipsing	Her X-1, XY Ari
	Atoll source	4U1608-52
	Z source	GX5-1, GX13+1
	Burster	MXB1730-335
	Black-hole candidate	GRS1915+105
	Cataclysmic variable	RS Oph, SS Cyg
	Symbiotic star	CH Cyg
Isolated neutron star		PSR B0656+14
Gamma-ray burst		GRB031203
Pulsar		Crab, Vela pulsar
Star	RS CVn	HR1099
	High-mass	η Carina
	O Supergiant	ζ Pup, τ Sco
	Dwarf	Algol, YY Gem
	Giant	Capella
	Supergiant	α Aqr, β Aqr
	White dwarf	HZ 43, Sirius
	Cepheids	δ Cep, β Dor
Star-forming regions		Orion nebula, h Per

with the motion of the planet or comet on the sky removed via post-processing. Such analyses are inherently more difficult than sources that do not move on the sky, and should not be used until after a few simpler objects have been analyzed.

A3.3   Calibration sources

The sources in Table A3.3 are taken from the International Astronomical Consortium for High Energy Calibration (IACHEC) website.[1] The IACHEC was formed to advance cross-calibration efforts between different missions, and has

---

[1] http://web.mit.edu/iachec

Table A3.2   *Common X-ray diffuse sources, by source type*

Category	Sub-category	Source name
Galaxy cluster	Bright	Perseus, M87
	high-z ($z > 1$)	3C186
	mid-z ($z \sim 0.2$)	Abell 665
	low-z ($z \approx 0.05$)	Abell 85, Abell 1795
Galaxy group		HCG 62, NGC 4410
Galaxy	Elliptical	NGC 720
	Spiral	NGC 253, NGC 4382
	Irregular	NGC1427A
	Starburst	M82
	Radio	Cen A, Cyg A
	Interacting	NGC4038
Supernova remnant	Shell-type	Cygnus Loop, Puppis
	Plerion	Crab Nebula, G21.5-0.9
	Mixed morphology	W44, W28
	Young	Cas A, Tycho
Solar System	Planet	Jupiter
	Comet	73P/Schwassmann–Wachmann

Table A3.3   *Common X-ray calibration sources, by source type*

Source name	Source type	Primary calibration use
HZ 43	White dwarf	High res. eff. area, $E < 1\,\mathrm{keV}$
Sirius B	White dwarf	High res. eff. area, $E < 1\,\mathrm{keV}$
GD 153	White dwarf	High res. eff. area, $E < 1\,\mathrm{keV}$
RX J1856.5-3753	Isolated neutron star	High res. eff. area, $E > 1\,\mathrm{keV}$
PSR B0656+14	Isolated neutron star	High res. eff. area, $E > 1\,\mathrm{keV}$
1E0102	Thermal SNR	Flux, spectrum
G21.5-0.9	Non-thermal SNR	Broad bandpass eff. area
Crab	Non-thermal SNR	Broad bandpass eff. area
Capella	Star	High res. wavelength, eff. area
PKS2155-304	AGN	On-axis eff. area from 0.1–10 keV

selected a small number of sources to use as recommended calibration targets. In general, these sources tend to be bright. Owing to their use as calibration sources, there may be a number of observations of each taken using unusual modes, in combination with regular normal observations.

Appendix 4

Major X-ray satellites

RANDALL K. SMITH

Tables A4.1, A4.2, and A4.3 list major X-ray satellites in order of their launch date along with some basic parameters about each. Table A4.1 includes all missions completed by 1980, and provides basic information about each including the detectors, energy bandpass, peak effective area, and FOV. None of these missions used focusing X-ray optics, relying instead upon collimators to select sources in a limited FOV.

Table A4.2 lists missions launched between 1980 and 1996 which are no longer operating, while Table A4.3 gives parameters for missions launched since 1996, which are still returning data as of 2010. Many of the missions in Tables A4.2 and A4.3 used X-ray optics and so more information is provided including the PSF and detector spectral resolution.

Our primary source for this information, especially for the Uhuru, ANS, Ariel-V, SAS-3, OSO-8, HEAO-1, the Einstein Observatory, EXOSAT, ASCA, RXTE, Swift, and Suzaku missions, was the HEASARC list of observatories.[1] In addition, the National Space Science Data Center spacecraft website[2] was invaluable. Both sites contain a large number of original source references. Information was also taken from the HEASARC calibration database CALDB in order to determine the peak effective area and spectral resolution. For the ROSAT HRI, some information also came from the SAO ROSAT site.[3] Specifications for Chandra,[4] XMM–Newton[5] and MAXI[6] came from their websites. Other sources are given in footnotes to the tables.

---

[1] http://heasarc.gsfc.nasa.gov/docs/observatories.html   [2] http://nssdc.gsfc.nasa.gov/nmc
[3] http://hea-www.harvard.edu/rosat/hricalrep.html   [4] http://cxc.harvard.edu
[5] http://xmm.esac.esa.int   [6] http://maxi.riken.jp

Table A4.1 *X-ray satellite missions completed before 1980*

Name	Launch year	Instrument	Bandpass/ keV	FOV	EA[1]/ cm^2	Detector[2]
Uhuru	1970–1973	0.5° coll.	1.7–18	0.52°	840	PC
		5° coll.	1.7–18	5.2°	840	PC
ANS[3]	1974–1976	SXX/USX	0.2–0.28	34′	144	PC
		SXX/UMX	1–7	34′ × 75′	45	PC
		HXX/LAD	1.5–30	10′ × 3°	60	PC
		HXX/BCS	1–4.2	—[5]	6	BCS
Ariel-V[4]	1974–1980	RMC	1.9–18	17°	n/a	PC
		SSI	0.9–18	All-sky	290	PC
		CPC	1.4–30	3.5°	100	PC
		LXCS	2–7	—[5]	60	BCS
		ST	26–1200	8°	n/a	CsI Sc
		ASM	3–6	All-sky	2	PC
SAS-3[6]	1975–1979	RMC	2–11	12°	178	PC
		SCPC	1–60	1° × 32°	75	PC
		TCPC	1–60	1.7°	80	PC
		LED	0.1–1	2.9°	8	PC
OSO-8[7]	1975–1978	GCXSE A	2–60	5°	263	PC
		GCXSE B	2–60	3°	37	PC
		GCXSE C	2–60	5°	237	PC
		HECXE	10–1000	5°	27.5	CsI(Na) Sc
		SXBRE	0.15–45	2.7°	52	PC
		GCXS	2–8	3°	11	BCS
HEAO-1	1977–1979	A1/LASS	0.25–25	1° × 4°	1900	PC
		A2/LED	0.15–3	3°	800	PC
		A2/MED	1.5–20	3°	800	PC
		A2/HED	2.5–60	3°	2400	PC
		A3/MC	0.9–13.3	30″	400	PC
		A4/LED	15–200	1.7°	200	PSc
		A4/MED	80–2000	17°	180	PSc
		A4/HED	120–10000	37°	100	PSc

[1] Characteristic value at the peak effective area or primary energy of interest
[2] PC = Proportional counter, BCS = Bragg Crystal Spectrometer,
Sc = Scintillator, PSc = Phoswich Scintillator
[3] No data available from this mission at HEASARC
[4] General information about Ariel-V (also known as Ariel 5) was taken
from Smith and Courtier (1976), with details on the CPC instrument
(also known as experiment C) from Bell Burnell and Chiappetti (1984)
and LXCS (aka experiment D) from Griffiths *et al.* (1976)
[5] Slitless spectrometer designed for point sources
[6] Parameters for SAS-3 came from Schnopper *et al.* (1976) for the RMC,
Buff *et al.* (1977) for the SCPC, Lewin *et al.* (1976) for the TCPC,
Hearn *et al.* (1976) for the LED
[7] Information about the OSO-8 GCXS is from Kestenbaum *et al.* (1976)

Table A4.2   *Completed X-ray satellite missions from 1980*

Name	Launch year	Instrument	Bandpass/ keV	PSF	EA[1]/ cm²	Resolution[1]/ eV @ keV
Einstein	1978–1981	IPC	0.4–4.0	1′	100	1600 @ 2
		HRI	0.15–3.0	2″	20	—[4] @ —
		SSS	0.5–4.5	—[2]	200	140 @ 2.1
		FPCS	0.42–2.6	—[2]	1.0	160 @ 1.0
		MPC	1.5–20	coll.[3]	667	1600 @ 6
Hakucho[5,9]	1979–1985	VSX	0.1–0.2	coll.[3]	155	n/a @ n/a
		SFX	1.5–30	coll.[3]	312	n/a @ n/a
		HDX	10–100	coll.[3]	57	n/a @ n/a
Tenma[6,9]	1983–1985	GSPC	2–60	coll.[3]	720	570 @ 6
		XFC	0.1–2	coll.[3]	15	n/a @ n/a
		TSM/HXT	2–25	coll.[3]	114	n/a @ n/a
		TSM/ZYT	1.5–25	coll.[3]	280	n/a @ n/a
		GBD	10–100	coll.[3]	14	—[4] @ —
EXOSAT	1983–1986	LE	0.05–2.0	18″	10	—[4] @ —
		ME	1–50	45′	1600	1200 @ 4.5
		GSPC	2–20	—[2]	100	760 @ 9
GINGA[7]	1987–1991	LAC	1.5–37	coll.[3]	4000	1080 @ 6
		ASM	1–20	coll.[3]	70	—[4] @ —
		GBD	1.5–500	All-sky	60	6000 @ 30
ROSAT	1990–1999	HRI	0.1–2.1	5″	95	—[4] @ —
		PSPC	0.1–2.4	20″	240	440 @ 1.2
ASCA	1993–2001	SIS	0.4–10	2.9′	440	90 @ 1.5
		GIS	0.7–10	2.9′	360	270 @ 2
BeppoSAX[8]	1996–2002	LECS	0.1–10	3.5′	22	310 @ 2
		MECS	1.3–10	1.2′	150	500 @ 5.8
		HPGSPC	4–120	coll.[3]	240	1200 @ 20
		PDS	15–300	coll.[3]	600	8300 @ 60

[1] Characteristic value at the peak effective area or primary energy of interest
[2] Slitless spectrometer designed for point sources
[3] X-rays are collimated, not focused
[4] No useful spectral resolution
[5] Information about Hakucho was taken from Kondo *et al.* (1981)
[6] Information about Tenma was taken from Tanaka *et al.* (1984)
[7] Information about Ginga was taken from Turner *et al.* (1989) and
Murakami *et al.* (1989)
[8] Information about BeppoSAX was taken from Boella *et al.* (1997)
[9] No data available from this mission at HEASARC

Table A4.3   *Operating (as of 2010) X-ray satellite missions*

Name	Launch year	Instrument	Bandpass/ keV	PSF	EA[1]/ cm^2	Resolution[1]/ eV @ keV
RXTE	1996–	PCA	2–60	coll.[2]	6500	3200 @ 10
		HEXTE	15–250	coll.[2]	1600	9500 @ 60
Chandra	1999–	ACIS-I	0.3–10	0.5″	580	140 @ 1.5
		ACIS-S	0.3–10	0.5″	670	100 @ 1.5
		HRC-I	0.06–10	0.5″	250	—[3] @ —
		HRC-S	0.06–10	0.5″	250	—[3] @ —
		HETG/HEG[5]	0.4–10	—[4]	67	2.3 @ 1.8
		HETG/MEG[5]	0.4–5	—[4]	145	3.0 @ 1.6
		LETG/ACIS	0.3–10	—[4]	12	0.7 @ 0.6
		LETG/HRC	0.07–10	—[4]	12	0.15 @ 0.19
XMM–	1999–	EPIC-MOS	0.15–12	14″	1100	94 @ 1.8
Newton		EPIC-pn	0.15–15	15″	1300	90 @ 1.5
		RGS	0.35–2.5	—[4]	120	1.2 @ 0.9
Swift	2004–	XRT	0.2–10	18″	110	77 @ 2.1
		BAT	15–150	coll.[2]	5240	5000 @ 30
Suzaku	2005–	XIS-FI	0.2–12	90″	400	98 @ 2.0
		XIS-BI	0.2–12	90″	400	110 @ 2.0
		PIN	10–30	coll.[2]	160	4000 @ 17
		GSO	30–600	coll.[2]	300	25000 @ 120
MAXI	2009–	GSC	2–30	coll.[2]	5000	1.1 @ 5.9
		SSC	0.5–10	coll.[2]	200	0.15 @ 5.9

[1] Characteristic value at the peak effective area or primary energy of interest
[2] X-rays are collimated, not focused
[3] No useful spectral resolution
[4] Slitless spectrometer designed for point sources
[5] With the ACIS detector in the focal plane

Appendix 5

Astrostatistics

ANETA SIEMIGINOWSKA

We briefly describe here a number of terms normally used by statisticians, with translations where appropriate into the terminology used in X-ray astrophysics; this information is taken from the CHASC jargon page at http://hea-www.harvard.edu/AstroStat/statjargon.html.

Background marginalization is integration of a background probability over uninteresting parameters.

Bias is a systematic difference between an estimated and a true value of a parameter.

Biased sample is a sample of objects selected from a population such that some objects are more likely to be included than others.

Bootstrap is a method for estimating parameter variance or other properties using an approximation to a distribution created by resampling the observed data themselves.

Cash statistic is a formulation of Poisson likelihood for a parametric model in X-ray astronomy.

Chi-square statistic is a statistic applied in X-ray astronomy which provides a measure of the goodness-of-fit. The name comes from the χ^2 distribution, however many of the "chi-square statistic" expressions do not follow the χ^2 distribution. Here are the most common expressions used:

> **model variance** $\chi^2 = (D - M)^2/M$
> **data variance** $\chi^2 = (D - M)^2/D$
> **iterative Primini approximation** $\chi_i^2 = (D - M_i)^2/M_{i-1}$, where i is the iteration fitting step.

Conditional distribution (or probability density) is the probability distribution of Y when X is known to be at a particular value and (X,Y) are variables in a joint distribution.

Confidence interval is a range of parameter values representing a quantifiable measure of plausibility (such as 95%, 99%, the level of confidence).

Data augmentation refers to methods for constructing iterative optimization or sampling algorithms via introduction of unobserved variables.

EM algorithm is a method to optimize a likelihood function in order to compute the maximum likelihood estimate (MLE) of an unknown parameter. It consists of two steps called Estimate and Maximize.

Estimator is a function of a random variable. Inserting data into the estimator provides an estimate (value) of the parameter. Estimators may be acquired from simple guessing or, more formally, via the maximum-likelihood method, Bayesian method, or other approaches.

Gibbs sampler is an MCMC sampler that constructs a Markov chain by dividing the set of unknown parameters into a number of groups and then simulating each group in turn, conditional on the current values of all the other groups.

Hypothesis testing is a statistical decision-making process. In general, to know the truth of the given hypothesis, some evidence (data) is collected with an assumption that these data are dependent upon the hypothesis truth. Therefore, summaries of the data set (statistics) should support the hypothesis. If the statistic is not consistent with the hypothesis, then the hypothesis can be rejected based on the data. There are many different test statistics that summarize data in order to make a statistical decision on a given hypothesis.

Likelihood quantifies the possibility of the observed data given a set of source model parameters.

Likelihood ratio (LR) test is a model selection method. It compares best-fit models, when one model is nested within the other.

Markov-Chain Monte Carlo (MCMC) is an algorithm used to generate simulations from a probability distribution. Because the simulations are generated with a Markov chain (the next sampled point depends only on the previous one), care must be taken to ensure that the chain has converged and to account for the auto-correlation in the simulation. Usually, more than one chain is run to assess the convergence. In Bayesian calculations, MCMC explores high-dimension posterior distributions in order to estimate unknown parameters and to construct error bars for these estimates.

Marginalization is integration over parameters. The name originates from historical methods of summing the columns in a table of parameters – on a margin of the table.

Martingale satisfies the following, $E(X_{n+1}|X_1, \ldots, X_n) = X_n$, where $X_1, \ldots, X_n$ are a sequence of random variables. In other words, the conditional expectation of X_{n+1}, given all past observations, only depends on the immediate previous observation.

Maximum likelihood estimator (MLE) of a given parameter is the function of the random variables where the likelihood is maximized.

Mean is an average value of a distribution. A number of different types are available such as:

arithmetic mean $\frac{1}{n}\sum_i X_i$
geometric mean $(\prod^n X_i)^{1/n}$
harmonic mean $n(\sum_i 1/X_i)^{-1}$

where n is a total number of random variables, X_i.

Median is a value (e.g. of a parameter) dividing the probability distribution into two equal parts, i.e. separating the higher half from the lower half.

Metropolis–Hastings sampler is an MCMC sampler that uses a rule to generate simulations and then uses an accept–reject criterion based on the likelihood function to correct the simulation to match the target probability distribution.

Mode of a probability distribution is the peak value of its probability density function (PDF). A distribution may have one or many modes, depending upon the shape of the PDF.

Model in statistics is a full description of the relationship between random variables and their probabilities. It is often a definition of a full likelihood function.

Model averaging refers to the process of estimating some quantity under each model and then averaging the estimates by the likelihood of each model.

Model selection is a statistical method to check and select the best model for the data. The best model is considered to minimize the distance to the true model, which, in general, is unknown. Often the selection is based on comparison of the likelihood functions.

Normal distribution is known as Gaussian distribution in astronomy (see Equation 7.2).

Poisson likelihood is a likelihood function with the Poisson distribution (see Equation 7.4).

Posterior distribution represents the updated knowledge about the unknown model parameters after observing the data and including all other information about the unknown parameters. Thus, the posterior distribution combines the information in the prior distribution and the likelihood (via Bayes theorem) and is a complete summary of the knowledge regarding the unknown model parameters.

Power, statistical of a statistical test is the probability of rejecting a false null hypothesis.

Principal components analysis (PCA) is a transformation of correlated variables into a number of uncorrelated variables called principal components, which are related to the original ones by an orthogonal transformation. The new variables are a linear combination of the original variables weighted by their contribution to that component.

Prior distribution is a probability distribution that quantifies knowledge about unknown quantities (e.g. model parameters) prior to observing the data.

Probability is for Bayesians a numerical measure of uncertainty and for Frequentists the fraction of events happening in a large number of identical trials.

Probability density function (PDF) (called "probability density" or "density") describes a relative likelihood for a random variable to occur at a certain value. It is usually calculated as an integral over a region. Its integral over the entire space is unity. The PDF is called the probability mass function for discrete distributions such as Poisson.

Random variable (r.v.) is a numerical value calculated from the outcome of a random experiment.

Sampling distribution is the probability distribution of a point estimate (e.g. a parameter value).

Skewness is a measure of the asymmetry of a random variable of a distribution.

Standard deviation (SD) is the square root of the variance.

Symbols Some commonly used symbols:

> $|$ – indicates conditional probability, e.g. $\text{Prob}(A|B)$ is read as the "the probability that A is true given that B is true."
>
> $\sim$ – usually written as $x \sim f(\ldots)$, denoting that the variable x is distributed as a function of the specified form. e.g. counts $\sim P_0(\lambda)$ or flux $\sim N(\mu, \sigma)$.
>
> $E(.)$ – expectation, or mean.
>
> λ – usually describes the Poisson intensity, unlike the astrophysical usage, where it is shorthand for wavelength.

Unbiased estimator is an estimator whose bias is equal to 0.

Variance is defined as $V(X) = E[(X - \mu)^2]$ and is a measure of the average fluctuation of a random variable X around its mean μ.

Appendix 6

Acronyms

The following tables list the acronyms used in this handbook, along with the chapter where they are defined or first used.

Table A6.1 *Acronyms*

Acronym	Ch.	Full expansion
ACE	8	Advanced composition explorer
ACIS	3	Advanced CCD Imaging Spectrometer (on Chandra)
ACIS-S	3	Advanced CCD Imaging Spectrometer – 6 × 1 spectroscopic array
ACIS-I	3	Advanced CCD Imaging Spectrometer - 2 × 2 imaging array
ADU	3	Analog-to-digital unit
AGN	I	Active galactic nucleus (or nuclei)
ANS	I	Astronomische Nederlandse Satelliet
APEC	5	Astrophysical Plasma Emission Code
ARF	4	Ancillary response file
ASCA	I	Advanced Satellite for Cosmology and Astrophysics
ASDC	6	ASI Data Center (at ESRIN)
ASM	6	All-Sky Monitor (on RXTE)
AtomDB	5	Atomic DataBase
AXAF	3	Advanced X-ray Astrophysics Facility (later renamed Chandra)
BAT	2	Burst Alert Telescope (on Swift)
BESSY	2	Berliner Elektronenspeicherring-Gesellschaft für Synchrotronstrahlung
BeppoSAX	I	Giuseppe "Beppo" Occhialini Satellite for X-ray Astronomy
CALDB	6	CALibration DataBase
CCD	I	Charge-coupled device
CCF	6	Current Calibration File
CDFN	6	Chandra Deep Field – North
CDFS	6	Chandra Deep Field – South
CfA	6	Harvard–Smithsonian Center for Astrophysics

Table A6.1   *(cont.)*

Acronym	Ch.	Full expansion
CHAMP	6	CHAndra Multiwavelength Project
CIAO	4	Chandra Interactive Analysis of Observations
COSPAR	I	COmmittee on SPAce Research
CTE	3	Charge-transfer efficiency
CTI	3	Charge-transfer inefficiency
CXC	I	Chandra X-ray Center
CZT	2	Cadmium zinc telluride
DARTS	6	Data ARchive and Transmission System (at ISAS)
DETX	4	DETector X-position
DETY	4	DETector Y-position
DN	3	Data number
EA	4	Effective-area map
EPIC	3	European Photon Imaging Camera (on XMM)
EPIC-pn	3	European Photon Imaging Camera p–n type
EPIC-MOS	3	European Photon Imaging Camera MOS type
ESA	I	European Space Agency
ESRIN	6	European Space Research INstitute
EUV	I	Extreme ultraviolet
EUVE	2	Extreme Ultraviolet Explorer
EXOSAT	I	European space agency's X-ray Observatory SATellite
FFT	5	Fast Fourier transform
FITS	4	Flexible Image Transport System
FOV	I	Field of view
FTOOLS	6	FITS TOOLS (part of HEAsoft)
FWHM	3	Full width at half-maximum
GEMS	I	Gravity and Extreme Magnetism SMEX
GIS	2	Gas Imaging Spectrometer (on ASCA)
GOODS	6	Great Observatories Origins Deep Survey
GSC	A4	Gas Slit Camera (on MAXI)
GSFC	6	NASA's Goddard Space Flight Center
GSPC	2	Gas scintillation proportional counter
GTI	4	Good time intervals
GUI	4	Graphical user interface
HEAsoft	4	High Energy Analysis software
HEAO	I	High Energy Astronomy Observatories
HEASARC	6	High Energy Astrophysics Science Archive Research Center
HEG	5	High Energy Grating (part of HETG on Chandra)
HETG	1	High Energy Transmission Grating (on Chandra)
HEXTE	2	High Energy X-ray Timing Experiment (on RXTE)
HPD	1	Half-power diameter
HPGSPC	2	High Pressure Gas Scintillation Proportional Counter (on BeppoSAX)
HRC	2	High Resolution Camera (on Chandra)
HRC-I	A4	High Resolution Camera – Imaging array
HRC-S	2	High Resolution Camera – Spectroscopic array
HRI	2	High Resolution Imager (on ROSAT)

(cont.)

Table A6.1   *(cont.)*

Acronym	Ch.	Full expansion
HST	6	Hubble Space Telescope
HXD	2	Hard X-ray Detector (on Suzaku)
IACHEC	A3	International Astronomical Consortium for High Energy Calibration
ICM	5	Intracluster medium
IDL	6	Interactive Data Language
INTEGRAL	5	INTErnational Gamma-Ray Astrophysics Laboratory
IPC	2	Imaging Proportional Counter (on Einstein Observatory)
IR	A2	Infrared
IRAF	6	Image Reduction and Analysis Facility
ISAS	I	Institute of Space and Astronautical Science
ISIS	6	Interactive Spectral Interpretation System
ISM	5	Interstellar medium
IXO	1	International X-ray Observatory
JAXA	I	Japan Aerospace eXploration Agency
KMC	2	Krystall MonoChromator (at BESSY)
LEDAS	6	LEicester Database and Archive Service
LETG	2	Low-energy transmission grating (on Chandra)
LHB	8	Local Hot Bubble
LR	7	Likelihood ratio
MAXI	I	Monitor of All-sky X-ray Image
MCMC	5	Markov Chain Monte Carlo
MCP	2	MicroChannel Plate
MEKAL	5	Mewe-Kaastra-Liedahl plasma emission code
MIT	I	Massachusetts Institute of Technology
MJD	4	Modified Julian date
MOS	3	Metal-oxide-semiconductor
NASA	I	National Aeronautics and Space Administration
NED	6	NASA/IPAC Extragalactic Database
NIST	2	National Institute of Standards and Technology
NuSTAR	I	Nuclear Spectroscopic Telescope ARray
ObsID	5	Observation IDentifier
OSO	I	Orbiting Solar Observatory
PCA	2	Proportional Counter Array (on RXTE)
PCU	2	Proportional Counter Unit (one part of the PCA)
PDS	2	Phoswich Detector System (on BeppoSAX)
PHA	4	Pulse-height amplitude
PI	4	PHA Invariant
PMT	2	Photomultiplier tube
PSF	4	Point spread function
PSPC	2	Position Sensitive Proportional Counter (on ROSAT)
PWN	8	Pulsar wind nebula
QE	2	Quantum efficiency
QPO	I	Quasi-periodic oscillation
RASS	6	ROSAT All-Sky Survey
RGS	1	Reflection grating spectrometer
RMF	2	Response matrix file

Table A6.1   *(cont.)*

Acronym	Ch.	Full expansion
RMS	3	Root mean square
ROSAT	I	Röentgen SATellite
RSP	4	ReSPonse matrix (ARF∗ RMF)
RXTE	I	Rossi X-ray Timing Explorer
SAA	3	South Atlantic Anomaly
SAS	4	Science Analysis System (for XMM–Newton)
SAS-3	2	Small Astronomical Satellite 3
SIMBAD	6	Set of Identifications, Measurements and Bibliography for Astronomical Data
SITAR	6	S-lang/Isis Timing Analysis Routines
SNR	5	Supernova remnant
SOC	I	Science Operations Centre (for XMM–Newton)
SPC	8	Soft-proton contamination
SPEX	6	SPEctral X-ray and UV modeling, analysis, and fitting
SQL	6	Structured Query Language
SSC	3	Solid-state Slit Camera (on MAXI)
SWCX	8	Solar-wind charge exchange
UT	4	Universal Time
UV	2	Ultraviolet
VO	6	Virtual Observatory
WCS	4	World Coordinate System
XIS	3	X-ray Imaging Spectrometer (on Suzaku)
XIS-BI	4	X-ray Imaging Spectrometer–Back Illuminated
XIS-FI	A4	X-Ray Imaging Spectrometer–Front Illuminated
XMM–Newton	I	X-ray Multi-Mirror Mission–Newton
XRS	3	X-Ray Spectrometer (on Suzaku)
XRT	3	X-Ray Telescope (on Swift)
XSA	6	XMM–Newton Science Archive

References

Alkhazov, G. D., 1970. *Nucl. Instrum. Meth. A*, **89**, 155.

Aschenbach, B., 1985. *Rep. Prog. Phys.*, **48**, 579.

Asplund, M., Grevesse, N., and Sauval, A. J., 2005. In *Cosmic Abundances as Records of Stellar Evolution and Nucleosynthesis in honor of David L. Lambert, ASP Conference Series Vol 336*, Barnes, T. G. and Frank, N. B. (eds.), Astronomical Society of the Pacific, San Francisco, CA, p. 25.

Avni, Y., 1976. *Astrophys. J.*, **210**, 642.

Babu, G. J. and Feigelson, E. D., 1996. *Astrostatistics*, Chapman and Hall, London.

Barthelmy, S. D. Barbier, L. M., Cummings, J. R., *et al.*, 2005. *Space Sci. Rev.*, **120**, 143.

Bell Burnell, S. J. and Chiappetti, L., 1984. *Astron. Astrophys. Suppl.*, **56**, 415.

Black, J. K., Baker, R. G., Deines-Jones, P., Hill, J. E., and Jahoda, K., 2007. *Nucl. Instrum. Meth. A*, **581**, 755.

Boella, G., Butler, R. C., Perola, G. C., *et al.*, 1997. *Astron. Astrophys. Suppl.*, **122**, 299.

Box, G. E. P. and Draper, N. R., 1987. *Empirical Model-Building and Response Surfaces*, Wiley, New York, NY.

Bradt, H., Swank, J., and Rothschild, R., 1991. *Advances Space. Res.*, **11**, 243.

Brandt, W. N. and Hasinger, G., 2005. *Ann. Rev. Astron. Astrophys.*, **43**, 827.

Brinkman, A., Aarts, H., den Boggende, A., *et al.*, 1998. In *Proceedings of the First XMM Workshop – Science with XMM*, published online at http://xmm.esac. esa.int/external/xmm_science/workshops/1st_workshop.

Broos, P. S., Townsley, L. K., Feigelson, E. D., *et al.*, 2010. *Astrophys. J.*, **714**, 1582.

Brown, G. V., Beiersdorfer, P., Liedahl, D. A., Widmann, K., and Kahn, S. M., 1998. *Astrophys. J.*, **502**, 1015.

Buff, J., Jernigan, G., Laufer, B., *et al.*, 1977. *Astrophys. J.*, **212**, 768.

Buote, D. A., 2000. *Mon. Not. Royal. Astron. Soc.*, **311**, 176.

Burrows, D. N., Hill, J. E., Nousek, J. A., *et al.*, 2000. *Proc. SPIE*, **4140**, 64.

Cappelluti, N., *et al.*, 2007. *Astrophys. J. Suppl.*, **172**, 341.

Cash, W., 1979. *Astrophys. J.*, **228**, 939.

Chen, L.-W., Fabian, A. C., and Gendreau, K. C., 1997. *Mon. Not. Royal. Astron. Soc.*, **285**, 449.

Cruddace, R. G., Hasinger, G. R., and Schmitt, J. H., 1988. *Eur. Southern Observatory Conference Workshop Proc.*, **28**, 177.

Davies, S. R., 1990. *Mon. Not. Royal. Astron. Soc.*, **244**, 93.

Davis, J. E., 2001a. *Astrophys. J.*, **548**, 1010.

Davis, J. E., 2001b. *Astrophys. J.*, **562**, 575.

Delgado-Martí, H., Levine, A. M., Pfahl, E., and Rappaport, S. A., 2001. *Astrophys. J.*, **546**, 455.

de Luca, A. and Molendi, S., 2004. *Astron. Astrophys.*, **419**, 837.

den Herder, J. W., Brinkman, A. C., Kahn, S. M., *et al.*, 2001. *Astron. Astrophys.*, **365**, L7.

Drake, G. W., 1988. *Can. J. Phys.*, **66**, 586.

Drake, J. J., Ratzlaff, P., Kashyap, V., *et al.*, 2006. *Proc. SPIE*, **6270**, 49.

Ebeling, H. and Wiedenmann, G., 1993. *Phys. Rev. E*, **47**, 704.

Erickson, G. W., 1977. *J. Phys. Chem. Ref. Data*, **6**, 831.

Evans, I. N., Primini, F. A., Glotfelty, K. J., *et al.*, 2010. *Astrophys. J. Suppl.*, **189**, 37.

Fabian, A. C., 1994. *Ann. Rev. Astron. Astrophys.*, **32**, 277.

Fano, U., 1947. *Phys. Rev.*, **72**, 26.

Ford, E. B., 2005. *Astron. J.*, **129**, 1706.

Foster, A., Smith, R. K., Brickhouse, N. S., Kallman, T. R., and Witthoeft, M. C., 2010. *Space Sci. Rev.*, **157**, 135.

Freeman, P. E., Kashyap, V., Rosner, R., and Lamb, D. Q., 2002. *Astrophys. J. Suppl.*, **138**, 185.

Frontera, F., Costa, E., dal Fiume, D., *et al.*, 1997. *Proc. SPIE*, **3114**, 206.

Fukazawa, Y., Makishima, K., Ebisawa, K., *et al.*, 1994. *Publ. Astron. Soc. Japan*, **46**, L55.

Gabriel, A. H., 1972. *Mon. Not. Royal. Astron. Soc.*, **160**, 99.

Gabriel, A. H. and Jordan, C., 1969. *Mon. Not. Royal. Astron. Soc.*, **145**, 241.

Garmire, G., Nousek, J., Apparao, K., *et al.*, 1992. *Astrophys. J.*, **399**, 694.

Garmire, G. P., Bautz, M. W., Ford, P. G., *et al.*, 2003. *Proc. SPIE*, **4851**, 28.

Giacconi, R. and Rossi, B., 1960. *J. Geophys. Res.*, **65**, 773.

Giacconi, R., Kellogg, E., Gorenstein, P., Gursky, H., and Tananbaum, H., 1971. *Astrophys. J.*, **165**, L27.

Giacconi, R., Branduardi, G., Briel, U., *et al.*, 1979. *Astrophys. J.*, **230**, 540.

Golub, L., 2003. *Rev. Sci. Instr.*, **74**, 4583.

Gregory, P. C. and Loredo, T. J., 1992. *Astrophys. J.*, **398**, 146.

Griffiths, R. E., Cooke, B. A., Peacock, A., Pounds, K. A., and Ricketts, M. J., 1976. *Mon. Not. Royal. Astron. Soc.*, **175**, 449.

Harrison, F. A., Boggs, S., Christensen, F., *et al.*, 2010. *Proc. SPIE*, **7732**, 21H.

Hearn, D. R., Richardson, J. A., Bradt, H. V. D., *et al.*, 1976. *Astrophys. J. Lett.*, **203**, L21.

Henke, B. L., Gullikson, E. M., and Davis, J. C., 1993. *Atomic Data Nuclear Data Tables*, **54**, 181.

Henley, D., Shelton, R. L., and Kuntz, K. D., 2007. *Astrophys. J.*, **661**, 304.

Herzberg, G., 1945. *Atomic Spectra and Atomic Structure*, Dover, New York.

Holt, S. S., Gotthelf, E. V., Tsunemi, H., and Negoro, H., 1994. *Publ. Astron. Soc. Japan*, **46**, L151.

Hubbell, J. H. and Seltzer, S. M., 2004. http://physics.nist.gov/PhysRefData/Xray MassCoef/cover.html.

Ikebe, Y., Böhringer, H., and Kitayama, T., 2004. *Astrophys. J.*, **611**, 175.

Jackson, J. D., 1998. *Classical Electrodynamics*, Wiley, New York, NY, Section 7.3.

Janesick, J. R., 2001. *Scientific Charge-Coupled Devices*, SPIE Publications, Bellingham, WA.

Johnson, W. R. and Soff, G., 1985. *Atomic Data Nuclear Data Tables*, **33**, 405.

Kaastra, J. S., 1992. *An X-Ray Spectral Code for Optically Thin Plasmas* (Internal SRON-Leiden Report, updated version 2.0).

Kashyap, V. L., van Dyk, D. A., Connors, A., *et al.*, 2010. *Astrophys. J.*, **719**, 900.

Kelley, R. L., Mitsuda, K., Allen, C. A., *et al.*, 2007. *Publ. Astron. Soc. Japan*, **59**, 77.

Kellogg, E., Murray, S., Giacconi, R., Tananbaum, H., and Gursky, H., 1973. *Astrophys. J.*, **185**, L13.

Kelly, B. C., Bechtold, J., and Siemiginowska, A., 2009. *Astrophys. J.*, **698**, 895.

Kelly, B. C., Sobolewska, M., and Siemiginowska, A., 2010. arXiv:1009.6011.

Kestenbaum, H. L., Cohen, G. G., Long, K. S., *et al.*, 1976. *Astrophys. J.*, **210**, 805.

Kondo, I., Inoue, H., Koyama, K., *et al.*, 1981. *Space Sci. Instrument.*, **5**, 211.

Koutroumpa, D., Lallement, R., Raymond, J. C., and Kharchenko, V., 2009. *Astrophys. J.*, **696**, 1517.

Koyama, K., Tsunemi, H., Dotani, T., *et al.*, 2007. *Publ. Astron. Soc. Japan*, **59**, 23.

Krause, M. O. and Oliver, J. H., 1979. *J. Phys. Chem. Ref. Data*, **8**, 329.

Kuntz, K. D. and Snowden, S. L., 2000. *Astrophys. J.*, **543**, 195.

Kuntz, K. D. and Snowden, S. L., 2008. *Astron. Astrophys.*, **478**, 575.

Lallement, R., 2004. *Astron. Astrophys.*, **418**, 143.

Lea, S. M., Silk, J., Kellogg, E., and Murray, S., 1973. *Astrophys. J. Lett.*, **184**, L105.

Leahy, D. A., Darbro, W., Elsner, R. F., Weisskopf, M. C., *et al.*, 1983a. *Astrophys. J.*, **266**, 160.

Leahy, D. A., Elsner, R. F., and Weisskopf, M. C., 1983b. *Astrophys. J.*, **272**, 256.

Lee, J. C., Xiang, J., Ravel, B., Kortright, J., and Flanagan, K., 2009. *Astrophys. J.*, **702**, 970.

Lewin, W. H. G., Doty, J., Clark, G. W., *et al.*, 1976. *Astrophys. J. Lett.*, **207**, L95.

Lewis, A. D., Buote, D. A., and Stocke, J. T., 2003. *Astrophys. J.*, **586**, 135.

Li, J., Kastner, J. H., Prigozhin, G. Y., Schulz, N. S., Feigelson, E. D., and Getman, K. V., 2004. *Astrophys. J.*, **610**, 1204.

Liedahl, D. A., Osterheld, A. L., and Goldstein, W. H., 1995. *Astrophys. J. Lett.*, **438**, 115.

Lisse, C. M., Dennerl, K., Englhauser, J., *et al.*, 1996. *Science*, **274**, 205.

Lutz, G., 2007. *Semiconductor Radiation Detectors: Device Physics*, Springer, Berlin.

McCammon, D., Burrows, D. N., Sanders, W. T., and Kraushaar, W. L., 1983. *Astrophys. J.*, **269**, 107.

Makino, F., 1987. *Astrophys. Lett.*, **25**, 223.

Manzo, G., Giarrusso, S., Santangelo, A., *et al.*, 1997. *Astron. Astrophys.*, **122**, 341.

Markevitch, M., Bautz, M. W., Biller, B., *et al.*, 2003. *Astrophys. J.*, **583**, 78.

Marshall, F. J. and Clark, G. W., 1984. *Astrophys. J.*, **287**, 633.

Matsuoka, M., *et al.*, 2009. *Publ. Astron. Soc. Japan*, **61**, 999.

Mewe, R., Lemen, J. R., and van den Oord, G. H. J., 1986. *Astron. Astrophys. Suppl.*, **65**, 511.

Muno, M. P., Bauer, F. E., Baganoff, F. K., *et al.*, 2009. *Astrophys. J.*, **181**, 110.

Murakami, T., Fujii, M., Hayashida, K., *et al.*, 1989. *Publ. Astron. Soc. Japan*, **41**, 405.

Nandra, K., George, I. M., Mushotzky, R. F., Turner, T. J., and Yaqoob, T., 1997. *Astrophys. J.*, **476**, 70.

Pearson, K., 1900. *Phil. Mag. Ser.*, **5**, **50**, 157.

Peterson, J. R., Kahn, S. M., Paerels, F. B. S., *et al.*, 2003. *Astrophys. J.*, **590**, 207.

Petre, R., Kuntz, K. P., and Shelton, R. L., 2002. *Astrophys. J.*, **579**, 404.

Pfeffermann, E., Briel, U. G., Hippmann, H., *et al.*, 1987. *SPIE Conf. Series*, **733**, 519.

Press, W. H., Teukolsky, S. A., Vetterling, W. T., and Flannery, B. R., 2007. *Numerical Recipes: The Art of Scientific Computing*, Third Edition, Cambridge University Press, Cambridge.

Protassov, R., van Dyk, D. A., Connors, A., Kashyap, V. L., and Siemiginowska, A., 2002. *Astrophys. J.*, **571**, 545.

Rana, V. R., Cook, W. R., Harrison, F. A., Mao, P. H., and Miyasaka, H., 2009. *Proc. SPIE*, **7435**, 2.

Raymond, J. C. and Smith, B. W., 1977. *Astrophys. J. Suppl.*, **35**, 419.

Robertson, I. P., Kuntz, K. D., Collier, M. R., *et al.*, 2009. In *The Local Bubble and Beyond II: Proceedings of the International Conference. AIP Conference Proceedings, Volume 1156*, Smith, R., Snowden, S., and Kuntz, K. (eds.), Springer, Berlin, p. 52.

Romaine, S. E., Basso, S., Bruni, R. J., *et al.*, 2004. *Proc. SPIE*, **5168**, 112.

Romer, A. K., Viana, P. T. P., Liddle, A. R., and Mann, R. G., 2001. *Astrophys. J.*, **547**, 594.

Rothschild, R. E., Blanco, P. R., Gruber, D. E., *et al.*, 1998. *Astrophys. J.*, **496**, 538.

Rots, A. and McDowell, J., 2008. http://fits.gsfc.nasa.gov/registry/region/region.pdf.

Scargle, J. D., 1983. *Astrophys. J.*, **263**, 835.

Scargle, J. D., 1998. *Astrophys. J.*, **504**, 405.

Schnopper, H. W., Delvaille, J. P., Epstein, A., *et al.*, 1976. *Astrophys. J. Lett.*, **210**, L75.

Seward, F. D., 2000. In *Allen's Astrophysical Quantities*, Cox, A. N. (ed.), Springer, New York, NY, p. 197.

Siemiginowska, A., Smith, R. K., Aldcroft, T. L., *et al.*, 2003. *Astrophys. J. Lett.*, **598**, L15.

Siemiginowska, A., Burke, D. J., Aldcroft, T. L., *et al.*, 2010. arXiv:1008.1739.

Smith, J. F. and Courtier, G. M., 1976. *Royal Soc. London Proc. Ser. A*, **350**, 421.

Smith, R. K., Brickhouse, N. S., Liedahl, D. A., and Raymond, J. C., 2001. *Astrophys. J. Lett.*, **556**, L91.

Smith, R. K. (ed.), 2005. *X-ray Diagnostics of Astrophysical Plasmas: Theory, Experiment, and Observation, AIP Conference Proceedings Volume 774*, Springer, Berlin.

Smith, R. K., Bautz, M. W., Edgar, R. J., *et al.*, 2007. *Publ. Astron. Soc. Japan*, **59**, S141.

Snowden, S. L., Egger, R., Freyberg, M. J., *et al.*, 1997. *Astrophys. J.*, **485**, 125.

Snowden, S. L., Egger, R., Finkbeiner, D., Freyberg, M. J., and Plucinsky, P. P., 1998. *Astrophys. J.*, **493**, 715.

Snowden, S. L., Mushotzky, R. F., Kuntz, K. D., and Davis, D. S., 2008. *Astron. Astrophys.*, **478**, 615.

Strüder, L., Briel, U., Dennerl, K., *et al.*, 2001. *Astron. Astrophys.*, **365**, L18.

Swank, J., Kallman, T., Jahoda, K., *et al.*, 2009. In *X-ray Polarimetry: A New Window in Astrophysics*, Bellazzini, R., Costa, E., Matt, G., and Tagliaferri, G. (eds.), Cambridge University Press, Cambridge, p. 251.

Sze, S. M., 2002. *Semiconductor Devices: Physics and Technology*, Wiley, Hoboken, NJ.

Takahashi, T., Abe, K., Endo, M., *et al.*, 2007. *Publ. Astron. Soc. Japan*, **59**, S35.

Tanaka, Y., Fujii, M., Inoue, H., *et al.*, 1984. *Publ. Astron. Soc. Japan*, **36**, 641.

Tanaka, Y., Inoue, H., and Holt, S. S., 1994. *Publ. Astron. Soc. Japan*, **46**, L37.

Townsley, L. K., Broos, P. S., Feigelson, E. D., Garmire, G. P., and Getman, K. V., 2006. *Astron. J.*, **131**, 2164.

Turner, M. J. L., Smith, A., and Zimmermann, H. U., 1981. *Space Sci. Rev.*, **30**, 513.

Turner, M. J. L., Thomas, H. D., Patchett, B. E., *et al.*, 1989. *Publ. Astron. Soc. Japan*, **41**, 345.

Turner, M. J. L., Abbey, A., Arnaud, M., *et al.*, 2001. *Astron. Astrophys.*, **365**, L27.

Ueno, S., Mushotzky, R. F., Koyama, K., *et al.*, 1994. *Publ. Astron. Soc. Japan*, **46**, L71.

Uttley, P., McHardy, I. M., and Vaughan, S., 2005. *Mon. Not. Royal. Astron. Soc.*, **359**, 345.

van Dyk, D. A., Connors, A., Kashyap, V. L., and Siemiginowska, A., 2001. *Astrophys. J.*, **548**, 224.

Vaughan, B. A., van der Klis, M., Wood, K. S., *et al.*, 1994. *Astrophys. J.*, **435**, 362.

Verde, L., Peiris, H. V., Spergel, D. N., *et al.*, 2003. *Astrophys. J. Suppl.*, **148**, 195.

Verner, D. A., Ferland, G. J., Korista, K. T., and Yakovlev, D. G., 1996. *Astrophys. J.*, **465**, 487.

Voges, W., Aschenbach, B., Boller, Th., *et al.*, 1999. *Astron. Astrophys.*, **349**, 389.

Wall, J. V. and Jenkins, C. R., 2003. *Practical Statistics for Astronomers*, Cambridge University Press, Cambridge.

Wang, Q. D., Gotthelf, E. V., and Lang, C. C., 2002. *Nature*, **415**, 148.

Watson, M. G., Schröder, A. C., Fyfe, D., *et al.*, 2009. *Astron. Astrophys.*, **493**, 339.

Wells, D. C., Greisen, E. W., and Harten, R. H., 1981. *Astron. Astrophys. Suppl.*, **44**, 363.

Westbrook, O. W., Evans, N. R., Wolk, S. J., *et al.*, 2008. *Astrophys. J. Suppl.*, **176**, 218.

White, N. E. and Peacock, A., 1988. In *Astronomy with EXOSAT*, Pallavicini, R. and White, N. E. (eds.), Consiglio Nazionale delle Ricerche, Societa Astronomica Italiana, Rome, p. 7.

White, D. A., Fabian, A. C., Johnstone, R. M., Mushotzky, R. F., and Arnaud, K. A., 1991. *Mon. Not. Royal. Astron. Soc.*, **252**, 72.

Wilks, W. R., Richardson, S., and Spiegelhalter, R. J., 1996. *Markov Chain Monte Carlo in Practice*, Chapman and Hall/CRC Interdisciplinary Statistics, London.

Wilms, J., Allen, A., and McCray, R., 2000. *Astrophys. J.*, **542**, 914.

Wolter, H., 1952a. *Ann. Physik*, **10**, 94.

Wolter, H., 1952b. *Ann. Physik*, **10**, 286.

Wright, E. L., 2006. *Publ. Astron. Soc. Pac.*, **118**, 1711.

Zhao, P. and Van Speybroeck, L. P., 2003. *Proc. SPIE*, **4851**, 124.

Index